Parametric Integer Programming

University of Missouri Studies LXVII

Parametric Integer Programming

Robert M. Nauss

University of Missouri Press
Columbia & London

Library of Congress Cataloging in Publication Data

Nauss, Robert M., 1947-
    Parametric Integer Programming.

    (University of Missouri studies; v. 65, 2)
    1. Integer programming. I. Title. II. Series:
University of Missouri—Columbia. University of
Missouri studies; v. 65, 2.
T57.7.N38   519.7′7   77-20207
ISBN 0-8262-0250-0

Acknowledgments

I wish to express my gratitude to Prof. Arthur M. Geoffrion. He was instrumental in laying the groundwork for much of what appears in this work. His criticisms and suggestions in the course of our many discussions were both insightful and inspiring.

This research was partially supported by the National Science Foundation under Grant GP-36090X and the Office of Naval Research under Contract N00014-69-A-0200-4042.

R. M. N.
St. Louis, Missouri
March 1977

To Mom and Dad,
who helped make this work possible

Contents

List of Tables

Parametric Integer Programming

I. Introduction

A parametric integer linear program (PILP) may be defined as a family of closely related integer linear programs (ILP). Parametric linear programming (PLP) theory is firmly entrenched, and a parametric capability is provided in most commercial linear programming (LP) packages. PILP, on the other hand, is a virgin field. This is natural since until recently methods for solving ILPs were not efficient. However, in the past few years the state of the art for ILP has developed to such an extent that research on PILP solution techniques may be undertaken with some optimism.

PLP is traditionally thought of as varying a scalar parameter continuously over a specified range, resulting in a continuum of objective functions or of right-hand sides (resource allocations). However, it is not used as extensively as one might think because the solution output generally contains much more information than management wants or needs, and partly because very small changes in the data are often not of interest. Rather, management generally desires the solution to a finite number of revised models. In the interest of solving the entire collection of problems efficiently, the analyst generally uses the optimal LP basis for one problem as an initial basis for a revised problem.

Analogously, while continuous parameterization is of some interest in PILP, the definition of PILP should be expanded to include finite parameterizations. Specifically, one may vary a parameter over a fixed number of points instead of over a continuous range, thus resulting in a finite number of objective functions or right-hand sides. An example is a capital budgeting problem modeled as an ILP. Since the precise cost of capital is rarely known (generally an educated guess is made), a logical approach would be to generate an objective function for each of a series of estimated costs of capital. Using this finite set of objective functions, a finite number of ILPs (which heretofore had to be solved independently) is transformed into a PILP with a finite number of objective functions. Another type of parameterization that should be included in the definition is varying the objective function, the right-hand side, and/or the constraint coefficients simultaneously. Varying the constraint coefficients could also be defined to include adding or deleting variables and constraints. PILP, then, may be divided into three broad categories:

1. Parameterization over a finite number of points (including simultaneous changes in the objective function, right-hand side, and constraint coefficients)

2. One parameter varied continuously over a specified range

1

3. Two or more parameters varied continuously over specified ranges.

Due to the increased complexity of the third category, we shall address only the first two.

Mathematical representations of parameterizations for the two categories to be addressed are given below. Let x be an n-vector, b an m-vector, and A an $m \times n$ matrix.

Finite parameterization
For $k = 1, \ldots, K$ solve:

$$\min (c + f_k)x$$
$$(A + D_k)x \geqslant (b + r_k)$$
$$x_j \text{ integer}, j \varepsilon J$$

where f_k, D_k, r_k are conformable with c, A, b, respectively.

Continuous parameterizations
For $\forall \theta \varepsilon [0,1]$ solve:

$$\min (c + \theta f)x$$
$$Ax \geqslant b$$
$$x_j \text{ integer}, j \varepsilon J$$

and for $\forall \theta \varepsilon [0,1]$ solve:

$$\min cx$$
$$Ax \geqslant b + \theta r$$
$$x_j \text{ integer}, j \varepsilon J$$

where f, r are conformable with c, b, respectively, and where θ is a scalar. Of course, in the finite case the parameterization may be confined to only the objective function, or the right-hand side, or the constraint matrix.

At this point, some mention should be made of work that has already been done in this field. It appears that G. M. Roodman [1972, 1974] was the first to do any computational work in the area. Basically, he has devised a method for one-at-a-time (for example, one cost coefficient or one resource) sensitivity analysis utilizing the fathomed nodes in the branch and bound tree generated for the ILP. V. J. Bowman [1974] has addressed ILP sensitivity from a group theoretic point of view. However, the practicality of his method has not yet been demonstrated. H. Noltemeier [1970] has done some theoretical work in the area of ILP sensitivity, but he has not performed any computational studies. C. J. Piper and A. A. Zoltners [1976] have attacked the problem of solving closely related ILPs by finding a set of the k best feasible solutions to an ILP. Sufficiency tests are proposed, which if passed, assure that an optimal solution to a revised ILP remains in the set.

D. Klein and S. Holm [1977] have developed necessary and sufficient conditions for coefficient changes in an ILP for use when the ILP has been solved by a cutting plane algorithm, but computational experience has been limited. R. E. Marsten and T. L. Morin [1975] have melded some of the ideas presented in this work with dynamic programming principles for the PILP parameterized on the right-hand side. Preliminary computational experience appears to be promising. A recent paper by M. A. Radke [1975] is concerned with continuity theory in mixed integer programming. While it is essentially a theoretical treatise, it does prescribe methods for eliminating the bogey of discontinuity in some problems.

The plan of this work is as follows. In the remainder of this chapter the motivation for studying PILP is outlined by presenting types of analysis for which a PILP formulation may be effective. Basic solution methodologies for PILP are presented, and two rudimentary algorithms are given. Characteristics and properties of special parameterizations are given in Chapter II, and in Chapter III problem dependent techniques for improving algorithmic performance are set forth. Factors affecting the scheduling of solution priorities for the PILP are examined in Chapter IV, and three different priority schemes are presented. In chapters V, VI, and VII the ideas and results of chapters II through IV are applied to special problem classes including the 0–1 knapsack problem, the generalized assignment problem, and the capacitated facility location problem. Algorithms are stated, computational results cited, and conclusions drawn concerning the most efficient algorithms for each problem class. Finally in Chapter VIII an approach for the general PILP is proposed, and directions for future research are given.

We advise the reader that problem notations remain valid for individual chapters only.

A. The Need for PILP Algorithms

Oftentimes in practical ILP applications, finding an "optimal" solution to a model is not the only requirement. Managers may also be interested in solutions that are close to optimal, or they may want to know what happens if a certain change is made in the model. Some cost coefficients or right-hand sides may not be known with certainty, and hence the manager must know how the optimal solution behaves as these parameters are varied in the model. Rarely, then, is an optimal solution sufficient for the needs of management. In most applications, various types of analysis must be done, and many of them can be classified under the broad term, *PILP*. This is because they satisfy the criteria of being a family of closely related ILPs. Quite a few types of analysis are mentioned in A. M. Geoffrion [1974b]. These along with some others are given below.

Sensitivity analysis. When problem data in an ILP is not known precisely, point estimates must be used in the model. Varying this data over a range of estimates allows management to determine how sensitive an optimal solution and value are to changes in the data.

Shadow price analysis. In every linear programming solution, shadow prices are available from the final tableau. Those prices reflect the value of an extra unit of resource. Unfortunately, reliable shadow prices are not available in ILP. By varying a right-hand side coefficient and observing how the optimal value changes, it is possible to approximate the value of an extra unit of a particular resource.

Trade-off analysis. When two or more criteria are reflected in a model, management must know the trade-offs involved in balancing one criteria against another. For example, a trade-off curve between customer service and total cost in a distribution system depicts how a change in customer service affects total distribution cost. Such an analysis would be done by varying the customer service parameters in the model over a suitable range.

Continuity analysis [Radke, 1975]. In linear programming, continuity of the optimal solution value with respect to problem data is generally taken for granted. In integer programming, however, serious discontinuities are more likely to occur. The analyst is interested in finding sufficient conditions for which continuity holds. Failing this, he would like to identify points of serious discontinuity. If a discontinuity exists and it is an accurate representation of the real world system, he may want to alter the problem data in order to improve the objective value. If it is not an accurate representation, a reformulation of the model may be required so that it more accurately represents the real world system. The task of identifying discontinuities or verifying that there are none in the region of interest can often be accomplished by systematically varying problem data over some neighborhood. Thus, a number of closely related problems may be formulated as a PILP.

Contingency analysis. ILP models must make specific assumptions, and hence cannot handle all possible "states of nature." For instance, the uncertainty associated with major unlikely events might best be treated externally to the model via modified formulations of the model. Thus, a number of ILPs, each of which corresponds to some contingency, might be solved as a PILP.

Implementation priority analysis. Management needs to measure the importance of various components of an optimal solution. This information is used in assigning implementation priorities to various contemplated changes in the real world system. If some individual changes result in only marginal savings, they might be deferred until some later time, or possibly might not be implemented at all. Those changes that produce significant savings, however, might be implemented in the near future with an eye toward overall corporate restraints such as limited capital expenditures in a given period. Since each priority formulation is an ILP, it is clear that a number of such formulations may be solved as a PILP.

Each of these types of analysis demonstrates a need for solving closely related ILPs efficiently. The goal of this study is to satisfy this need.

B. Basic Methodologies for Solving the PILP

It is natural to look to ILP solution methodology for ideas on formulating a solution methodology for PILP. ILP solution techniques fall into three distinct categories: cutting plane, group theory, and branch and bound. R. S. Garfinkel and G. L. Nemhauser [1972] give a detailed account of each of these categories.

With one notable exception [G. T. Martin, 1963], cutting-plane techniques have not been overly successful in practice. Since cutting-plane methods are dual based, no feasible solutions are generated along the way to finding an optimal solution. This is a serious drawback since even though finite convergence is assured for many cutting-plane algorithms, no upper bound can be put on the number of iterations required to find an optimum. This lack of an upper bound on the convergence of the algorithms, coupled with the inability to systematically generate feasible solutions, generally makes cutting-plane methods unattractive in real-world ILP applications.

Group theoretic methods have been applied to pure ILPs for the most part. While some advances have been made, computational experience has shown that group methods are not as effective as the branch and bound approach, although they may be useful within a branch and bound framework.

Branch and bound methodology has come to the fore in ILP technology. Due to the inherent flexibility of this approach, problems with special structure can be solved efficiently by taking advantage of properties associated with the special structure. Furthermore, feasible solutions are often generated before optimality is proved so that if early termination is necessary, a good feasible solution will be available in most cases.

In the remainder of this chapter, some PILP analysis will be developed using cutting-plane methods. However, as would seem to be true in ILP, the most effective use of cutting planes appears to be incorporating them in a branch and bound approach. The bulk of the analysis consequently will be an outgrowth of ILP branch and bound methodology.

As a means for understanding the relationship between ILP and PILP solution techniques, two rudimentary algorithms for solving a PILP will be given. The first approach utilizes cutting planes while the second uses branch and bound. These algorithms should provide a springboard for the more detailed analysis in later chapters.

1. Cutting-Plane Approach

The main idea behind cutting-plane algorithms in ILP is to "cut off" portions of the linear programming feasible region while leaving the ILP feasible region (that is, the convex hull) untouched. The optimal linear programming solution thus approaches the optimal integer solution as the LP feasible region approaches that of the ILP feasible region. Computational experience has shown that the first few cuts are often very effective in removing large parts of the LP feasible region. However, later cuts become less and less effective, and progress toward an optimal ILP solution deteriorates.

This empirical property of cutting-plane algorithms might be exploited in PILP. Assume we have a PILP with a finite number of objective functions:

For $k = 1, \ldots, K$ solve:

$$\min \ (c + f_k)x$$
$$x \geq 0$$
$$(R_k) \qquad Ax \geq b$$
$$x_j \text{ integer, } j \, \varepsilon \, J$$

where c, f_k, A, b are all integer valued. It is easy to see that the feasible regions of each (R_k) are the same. If cuts are made only on the constraint set, then a cut that is valid for (R_1), say, is valid for all other (R_k). Of course, a cut made on the objective function may not be valid for all (R_k). Therefore, we assume in the following that such cuts are not allowed. Such a restriction could negate finite convergence, but the following approach could be easily modified to assure convergence for the price of some additional bookkeeping.

Since cutting-plane algorithms usually perform well during the first few iterations, a plausible approach for solving the PILP might be the following. We assume the feasible region is nonempty and bounded.

Solve (R_1) to LP optimality. If the optimal solution is not integer feasible, then cuts are added to (R_1) until they become "ineffective," and then retaining these cuts, (R_2) is solved to LP optimality. Cuts are then added to (R_2) until they become ineffective, and (R_3) is then solved to LP optimality, and so on. When cuts become ineffective for (R_K), return to (R_1), and continue the process. The hope is that cuts added to one problem will eliminate the long series of ineffective cuts for the other problem.

A cutting-plane algorithm for the PILP with a finite number of right-hand sides can also be constructed. Consider the problem:

For $k = 1, \ldots, K$ solve:

$$\min \ cx$$
$$x \geq 0$$
$$(Q_k) \qquad Ax \geq b + r_k$$
$$x_j \text{ integer, } j \, \varepsilon \, J.$$

In the general case, some constraints are tightened and others are relaxed as k varies, so that the feasible regions are not necessarily of the form $F(Q_1) \supseteq F(Q_2) \supseteq \ldots \supseteq F(Q_k)$. Up until now it was thought that cutting planes could not be used under such conditions, since a cut might be valid for one (Q_k)

but not for another. The following new result allows one to generate a valid cut for all K problems. See Klein and Holm [1977] for a more general proof and some preliminary computational results.

In the theorem we shall use the traditional LP notation of the simplex method. Let B be a basis, x_{B_i} be a basic variable under the basis B, NB be the index set of nonbasic variables, and $x_{B_i} = y_i - \sum_{j \varepsilon NB} \bar{a}_{ij} x_j$ where $y_i = (B^{-1}b)_i$, and $\bar{a}_{ij}$ is the updated constraint coefficient for the LP tableau associated with the basis B. We assume that (Q_k) is a pure integer program with c, r_k, A, b having all integer components for $k = 1, \ldots, K$. An analogous result holds for the mixed integer case as well.

Theorem 1. Given any basic (possibly primal infeasible) solution for (Q_k), the traditional Gomory cut, $\sum_{j \varepsilon NB} (\bar{a}_{ij} - [\bar{a}_{ij}]) x_j \geqslant y_i - [y_i]$ (using the ith row of the tableau as the source row), does not violate the convex hull of integer solutions for (Q_k).

Proof. Let x_{B_i} be a basic variable and suppose $y_i < 0$. Then $x_{B_i} = y_i - \sum_{j \varepsilon NB} \bar{a}_{ij} x_j \Rightarrow x_{B_i} + \sum_{j \varepsilon NB} [\bar{a}_{ij}] x_j \leqslant y_i$ since $x_j \geqslant 0 \ \forall j \varepsilon NB \Rightarrow x_{B_i} + \sum_{j \varepsilon NB} [\bar{a}_{ij}] x_j \leqslant [y_i]$ since both sides of the inequality must be integer. Then

$$x_{B_i} + \sum_{j \varepsilon NB} \bar{a}_{ij} x_j = y_i$$

$$- \left(x_{B_i} + \sum_{j \varepsilon NB} [\bar{a}_{ij}] x_j \leqslant [y_i] \right)$$

$$\overline{\sum_{j \varepsilon NB} (\bar{a}_{ij} - [\bar{a}_{ij}]) x_j \geqslant y_i - [y_i]}$$

or letting $f_{ij} = \bar{a}_{ij} - [\bar{a}_{ij}]$ and $f_i = y_i - [y_i]$,

we have $\sum_{j \varepsilon NB} f_{ij} x_j \geqslant f_i$. $\quad \|$

Note that the assumption that $y_i < 0$ did not affect the proof.

To use this result effectively, we proceed as follows. Set up the LP tableau for an initial basic feasible solution for (Q_1) under the basis B,

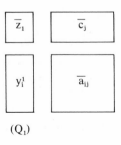

(Q_1)

where y_i^1 is the updated right-hand side and $\bar{z}_1$ is the associated objective function value. Then in columns directly to the left of the $[\bar{z}_1 \; y_i^1]$ column, add columns for initial (possibly infeasible) basic solutions for (Q_k), $k = 2, \ldots, K$ with respect to the initial basis B found for (Q_1).

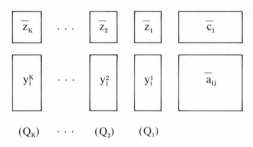

$$\begin{array}{ccccc} \boxed{\bar{z}_K} & \cdots & \boxed{\bar{z}_2} & \boxed{\bar{z}_1} & \boxed{\bar{c}_j} \\[2mm] \boxed{y_i^K} & \cdots & \boxed{y_i^2} & \boxed{y_i^1} & \boxed{\bar{a}_{ij}} \\[2mm] (Q_K) & \cdots & (Q_2) & (Q_1) \end{array}$$

Now solve the LP for (Q_1) updating the tableau from iteration to iteration in the usual manner with the extra columns for $k = 2, \ldots, K$ being updated according to the same rules as for $k = 1$. Assume we have solved the LP to optimality for (Q_1) and the solution is not integer. Then we may add the cut (choosing a source row i): $\sum_{j \in NB} f_{ij} x_j \geq f_i^1$. But by Theorem 1 we may add the cut $\sum_{j \in NB} f_{ij} x_j \geq f_i^k$ (using the same source row) for (Q_k), $k = 2, \ldots, K$. Note that $\sum_{j \in NB} f_{ij} x_j$ is the same for all k, so that only one row must be added to the tableau. This row will be $\sum_{j \in NB} f_{ij} x_j \geq (f_i^1, f_i^2, \ldots, f_i^K)$. The expanded tableau is:

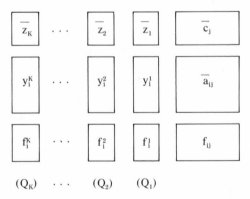

$$\begin{array}{ccccc} \boxed{\bar{z}_K} & \cdots & \boxed{\bar{z}_2} & \boxed{\bar{z}_1} & \boxed{\bar{c}_j} \\[2mm] \boxed{y_i^K} & \cdots & \boxed{y_i^2} & \boxed{y_i^1} & \boxed{\bar{a}_{ij}} \\[2mm] \boxed{f_i^K} & \cdots & \boxed{f_i^2} & \boxed{f_i^1} & \boxed{f_{ij}} \\[2mm] (Q_K) & \cdots & (Q_2) & (Q_1) \end{array}$$

So then for any basis B we may add a cut that is valid for each (Q_k). Unfortunately, a cut guarantees only to cut off the current solution, and not to cut off any feasible integer solutions. So if the current basic solution is infeasible for some (Q_k), it is possible that none of the feasible LP region of (Q_k) will be cut off. However, if the current infeasible solution is "close" to the feasible region, it is more likely that some of the feasible LP region will be cut off.

It follows that the cutting-plane approach given for the finite objective function PILP can be used also for this finite right-hand side PILP assuming that for each (Q_k) the feasible region is nonempty and bounded. By making simple changes to allow for unbounded or empty feasible regions, this assumption may be dropped.

2. *Branch and Bound Approach*

We now turn our attention to the method of branch and bound. Due to the inherent flexibility of the approach, it is not surprising that the general ILP branch and bound approach may be extended in a straightforward manner to the PILP. We shall give a rudimentary ILP algorithm and then show how it can be generalized to the PILP.

Consider the ILP:

$$\min cx$$
$$(P) \qquad Ax \geqslant b$$
$$x_j \text{ integer}, j \, \varepsilon \, J.$$

Let $v(\cdot)$ be the optimal value of $(\cdot)$. A rudimentary branch and bound approach for (P) is [Geoffrion and Marsten, 1972]:

1. Initialize the candidate list to consist of (P) and set z^* to ∞.

2. Stop if the candidate list is empty: if z^* is finite, then the solution x^* associated with z^* is optimal in (P); otherwise (P) has no feasible solution.

3. Select one of the candidate problems to become the current candidate problem (CP).

4. Solve a relaxation of (CP), namely (CP_R).

5. If $F(CP_R) = \phi$ or $v(CP_R) \geqslant z^*$, then go to 2.

6. If the optimal solution to (CP_R) is integer feasible, set $z^* = v(CP_R)$, set x^* to the optimal solution of (CP_R), and go to 2.

7. Separate (CP) into two simpler problems such that the union of their feasible regions is the feasible region of (CP). Add these two problems to the candidate list and go to 3.

We now consider the PILP.

For $k = 1, \ldots, K$ solve:

$$\min (c + f_k)x$$
$$(P_k) \qquad (A + D_k)x \geqslant b + r_k$$
$$x_j \text{ integer}, j \, \varepsilon \, J.$$

Candidate problems will be denoted by (P_{k,R_i}). The subscript k refers to the particular problem in the PILP, while R_i refers to a particular restriction placed on (P_k) as a result of separation. An example of such a restriction would be to append the constraint $x_1 = 1$ to (P_k). Define R_0 to be the null restriction, so that $(P_{k,R_0}) = (P_k)$ $\forall\, k = 1, \ldots, K$.

A rudimentary branch and bound approach for the PILP is:

1. Initialize the candidate list to consist of (P_{1,R_0}), $P_{2,R_0})$, $\ldots$, (P_{K,R_0}) and set z_k^* to ∞ for $k = 1, \ldots, K$.

2. Stop if the candidate list is empty: for each $k = 1, \ldots, K$, if z_k^* is finite, then x_k^* is optimal in (P_k); otherwise (P_k) has no feasible solution.

3. Select some subset, S, of the candidate list such that each member of S has the same R_i. For $s \, \varepsilon \, S$ denote the corresponding candidate problem by $(CP_{k,R_i})_s$.

4. For each $s \, \varepsilon \, S$, solve a relaxation, say, $(\overline{CP}_{k,R_i})_s$, of $(CP_{k,R_i})_s$.

5. For each $s \, \varepsilon \, S$, if either $F(\overline{CP}_{k,R_i})_s = \phi$ or $v((\overline{CP}_{k,R_i})_s) \geqslant z_k^*$, then delete s from S. If S is empty, go to 2.

6. For each $s \, \varepsilon \, S$, if an optimal solution to $(\overline{CP}_{k,R_i})_s$ is integer feasible, set $z_k^* = v((\overline{CP}_{k,R_i})_s)$ and x_k^* to the optimal solution of $(\overline{CP}_{k,R_i})_s$ and delete s from S. If S is empty, go to 2.

7. For each $s \, \varepsilon \, S$, separate $(CP_{k,R_i})_s$ into two simpler problems such that the union of the feasible regions is the feasible region of $(CP_{k,R_i})_s$. Update R_i to reflect this added restriction, and go to 3.

Through the choice of the subset S of candidate problems in step 3, flexibility is permitted in determining the order (if any) in which the individual (P_k)s are to be solved. By restricting the choice of S to those candidate problems with the same R_i, it may be possible to utilize parametric reoptimization techniques in step 4. This is the main reason for allowing more than one candidate problem at a time to be selected from the candidate list in step 3. In steps 5 and 6 if fathoming occurs for some $s \, \varepsilon \, S$, then s may be deleted from S since any further restrictions of $(CP_{k,R_i})_s$ are not of interest. In step 7 separation occurs for all remaining $s \, \varepsilon \, S$. To capitalize fully on parametric reoptimization techniques in step 4, it may be advisable to invoke identical separations for all $s \, \varepsilon \, S$. However, this is not required. Basically, then, the choice of S in step 3 and the choice of relaxation in step 4 allows fathoming and separation machinery to be applied in such a way as to limit the number of relaxations to be solved and to have control over the creation of new candidate problems.

Armed with these rudimentary approaches for PILP our next task is to identify and catalog salient characteristics and properties of special types of parameterizations that may be of use in formulating more sophisticated algorithms. This is the topic of the next chapter.

II. Characteristics and Properties of Particular Parameterizations

Aside from characteristics of specific problem classes, a PILP possesses other characteristics and properties that may influence the solution strategies to be used in an efficient algorithm. These attributes generally depend on the type of analysis that is undertaken, for example, trade-off, sensitivity, and priority. A representative checklist of attributes is:

1. Are all the ILP problems that make up the PILP known in advance? An example of where this may not be the case is in a priority analysis. This analysis depends in a sequential fashion on the optimal solutions to selected ILPs; generally, it is impossible to state explicitly in advance all of the ILPs that are to be solved.

2. How many ILP problems make up the PILP? Are there just a few, a dozen, a continuum, or is the number unknown?

3. How are the individual ILP problems related to one another? Do they differ only in the objective function or only in the right-hand side? Does the constraint matrix change? Are variables added or deleted? Are constraints added or deleted? Is there a continuous parameter of change? Are the feasible regions or the objective functions changing monotonely?

4. What optimality tolerance is required? Is ε-optimality or "beating some threshold" the criterion? Is the threshold value known in advance for each problem or does it depend on intermediate solution results?

Associated with some of these attributes are properties that may be used to advantage in algorithmic design. Some of these properties allow the transformation of a given PILP into a hopefully simpler PILP. Others deal with the behavior of the optimal solution and value for special types of parameterizations.

A. Transformation of a Continuous Pure PILP to a Finite PILP

Consider the *pure integer* PILP:

For $\forall \, \theta \, \varepsilon \, [0,1]$ solve:

$$
\begin{aligned}
&\min cx \\
(T_\theta) \qquad &Ax \geqslant b + \theta r \\
&x \text{ integer}
\end{aligned}
$$

where x is an n-vector and b is an m-vector. We assume that each component of c, b, r, and each entry of A is integer valued. The elementary result that follows [compare Noltemeier, 1970] shows that it is possible to transform the *continuum* of problems over $\forall\, \theta\, \varepsilon\, [0,1]$ into an equivalent PILP with a *finite* number of right-hand sides.

Theorem 2. The PILP, $(T_\theta)\ \forall\, \theta\, \varepsilon\, [0,1]$, may be transformed into an equivalent PILP of the form:

For $k = 1, \ldots, K$ solve:

$$\begin{aligned} &\min cx \\ (T_k) \qquad &Ax \geqslant b + t_k \\ &x \text{ integer} \end{aligned}$$

where t_k is an integer-valued vector conformable with b for $k = 1, \ldots, K$.

Proof. Since all entries are integer and x is required to be integer, then for a given $\hat{\theta}\, \varepsilon\, [0,1]$ the vector $b + \hat{\theta} r$ may be replaced by the vector $<b + \hat{\theta} r>$ (where $<\cdot>$ denotes the smallest integer greater than or equal to $\cdot$, component by component). It follows that it is sufficient to solve the PILP for those $\theta\, \varepsilon\, [0,1]$ such that θr_i is integer $(r_i \neq 0)$ for some $i\, \varepsilon\, \{1, \ldots, m\}$. But only a finite number of values of $\theta\, \varepsilon\, [0,1]$ satisfies this property, since each component of r is finite. Hence the PILP may be transformed to the problem in the theorem statement. ‖

The actual transformation may be done in the following way. For each component $r_i \neq 0$ of r write down the values of $\theta\, \varepsilon\, [0,1]$ for which r_i is an integer. These values are 0, $(1/|r_i|)$, $(2/|r_i|)$, $\ldots$, $(|r_i - 1|/|r_i|)$, 1. Place these values in a set R_i. Then $H = \bigcup_{i=1}^{m} R_i$ is the set of values of θ for which some θr_i is an integer. Let K be the cardinality of H, and let θ_k be an element of H. Then an equivalent PILP is:

For $k = 1, \ldots, K$ solve:

$$\begin{aligned} &\min cx \\ &Ax \geqslant b + \theta_k r \\ &x \text{ integer.} \end{aligned}$$

Letting $t_k = \theta_k r$ we have the result stated in the theorem.

We define the *range* of a continuous right-hand side parameterization to be $[b, b + r]$ where the closed interval is taken component by component over $i = 1, \ldots, m$. The next result shows that it is possible to expand the range of parameterization of $(T_\theta)\ \forall\, \theta\, \varepsilon\, [0,1]$ by increasing certain $|r_i|$, and through this adjustment to transform the continuum of problems into a finite PILP with fewer right-hand sides than under the transformation associated with the original (smaller) range of parameterization. We note that such an expansion of the parameterization does change the original problem. However, if certain r_i's have been chosen arbitrarily (such as $r_i = [b_i \cdot \text{constant}]$), then small perturbations in r_i may be reasonable in light of the potential reduction in the number of problems to be solved.

Theorem 3. For (T_θ) $\forall \varepsilon [0,1]$ let K be the number of right-hand sides resulting from the transformation of Theorem 2. It is possible to modify r such that the range of the parameterization is increased and the resulting finite transformation has K' right-hand sides where $K' \leqslant K$.

Proof. Let $M^* = \max_i |r_i|$. Find the prime factors $p_1, p_2, \ldots, p_l$ of M^*. So $M^* = p_1 \cdot p_2 \cdot \ldots \cdot p_l$. For $\forall i$ such that $|r_i| < M^*$ do the following:
1. If $r_i > 0$, find $r_i' = \min \{\bar{M} | \bar{M} \geqslant r_i$ and $\bar{M} = \prod_{j \varepsilon Q_i} p_j\}$ where Q_i is some subset of $\{1, 2, \ldots, l\}$, the index set of prime factors of M^*.
2. If $r_i < 0$, find $r_i' = \max \{\bar{M} | \bar{M} \leqslant r_i$ and $\bar{M} = - \prod_{j \varepsilon Q_i} p_j\}$.

Then $b + \theta r'$ covers a larger range than $b + \theta r$ for $\forall \theta \varepsilon [0,1]$, and the only values of θ for which $\theta r_i'$ (for some i) may be integer are $0, (1/M^*), (2/M^*), \ldots, (M^* - 1/M^*), 1$. Since $M^* + 1$ is the minimum value that K could take on, the new transformation has $K' \leqslant K$. ‖

An example should make this proof clearer. Assume (T_θ) $\forall \theta \varepsilon [0,1]$ has only two constraints exclusive of integrality requirements. Let the two constraints be:

$$\sum_j a_{1j} x_j \geqslant b_1 + \theta r_1 = b_1 + \theta \cdot 9$$

$$\sum_j a_{2j} x_j \geqslant b_2 + \theta r_2 = b_2 + \theta \cdot 2.$$

Now the values of $\theta \varepsilon [0,1]$ for which θr_1 is an integer are $\{0, \frac{1}{9}, \ldots, \frac{8}{9}, 1\} = R_1$, and for θr_2 are $\{0, \frac{1}{2}, 1\} = R_2$. Then $R_1 \cup R_2 = \{0, \frac{1}{9}, \frac{2}{9}, \frac{3}{9}, \frac{4}{9}, \frac{1}{2}, \frac{5}{9}, \frac{6}{9}, \frac{7}{9}, \frac{8}{9}, 1\}$ and $K = 11$. Now apply the theorem: $M^* = \max \{|9|, |2|\} = 9$, so $M^* = 3 \cdot 3 \cdot 1 = p_1 \cdot p_2 \cdot p_3$. Since $r_2 > 0$, we find $r_2' = \min \{\bar{M} | \bar{M} \geqslant r_2 = 2$ and $\bar{M} = \prod_{j \varepsilon Q_2} p_j\}$. A set Q_2 that satisfies the equation is $Q_2 = \{2,3\}$. So $\bar{M} = \prod_{j \varepsilon [2,3]} p_j = 3 \cdot 1 = 3$ and $r_2' = 3$. The new system of constraints is:

$$\sum_j a_{1j} x_j \geqslant b_1 + \theta \cdot 9$$

$$\sum_j a_{2j} x_j \geqslant b_2 + \theta \cdot 3$$

which for $\theta \varepsilon [0,1]$ covers a larger range. Furthermore, $R_1' \cup R_2' = \{0, \frac{1}{9}, \frac{2}{9}, \frac{3}{9}, \frac{4}{9}, \frac{5}{9}, \frac{6}{9}, \frac{7}{9}, \frac{8}{9}, 1\}$ and $K' = 10$.

Of course if M^* is prime, then $r_i' = M^* \; \forall \; i = 1, \ldots, m$, which may result in too large a range and may not be of interest. However, this type of transformation may be done over some subset of r_i, or over some restricted range for r_i' such that the resultant range is of interest.

B. Theoretical Properties for Right-Hand Side and Constraint Matrix Parameterizations

Consider the PILP:

For $k = 1, \ldots, K$ solve:

$$\text{min } cx$$
$$(S_k) \qquad (A + D_k)x \geq b + r_k$$
$$x_j \text{ integer, } j \; \varepsilon \; J.$$

Note that this parameterization allows simultaneous changes in the constraint matrix and the right-hand side. Let $F(S_k)$ be the respective feasible regions, $v(S_k)$ be the optimal solution values, and x_k^* be an optimal solution for (S_k).

Definition. A PILP is said to be *monotone* if $F(S_1) \supseteq F(S_2) \supseteq \ldots \supseteq F(S_K)$.

Theorem 4. Let the PILP be monotone. If x_k^* is optimal for (S_k) and $x_k^* \; \varepsilon \; F(S_{k+1})$, then x_k^* is optimal for (S_{k+1}) also.

Proof. Since $F(S_{k+1}) \subseteq F(S_k)$, $v(S_{k+1}) \geq v(S_k)$. Now $x_k^* \; \varepsilon \; F(S_{k+1})$ so $v(S_{k+1}) = v(S_k)$ and x_k^* is optimal for (S_{k+1}). ‖

It is clear that if a PILP (or even some subset of the PILP) is monotone, and the individual problems are solved in the order of decreasing feasible regions, then some (S_k) may not have to be solved. This occurs if $x_k^* \; \varepsilon \; F(S_{k+1})$.

We mention in passing that a well-known PLP result for a continuous parameterization of the right-hand side does not hold for PILP. Consider the PLP:

For $\forall \; \theta \; \varepsilon \; [0,1]$ solve:

$$\text{min } cx$$
$$x \varepsilon X$$
$$(\bar{H}_\theta) \qquad Ax \geq b + \theta r$$

where X is a polyhedral containing upper and lower bounds on all variables, thus assuring a bounded feasible region. It is well known that the optimal solution value $v(\bar{H}_\theta)$ is piecewise linear, continuous, and convex. Unfortunately, this does not hold for the corresponding PILP. However, we do have the following simple result.

Consider the PILP:

For $\forall \, \theta \, \varepsilon \, [0,1]$ solve:

$$\min_{x \varepsilon X} cx$$

(H_θ)
$$Ax \geqslant b + \theta r$$
$$x_j \text{ integer}, j \, \varepsilon \, J.$$

Let the PILP be monotone. That is, if $0 \leqslant \theta_1 \leqslant \theta_2 \leqslant 1$, then $F(H_{\theta_1}) \supseteq F(H_{\theta_2})$. Also assume that $F(H_1) \neq \phi$.

Theorem 5. The optimal solution value $v(H_\theta)$ is nondecreasing (and hence quasiconvex) on $[0,1]$.

Proof. Immediate. ‖

Note that $v(H_\theta)$ will be piecewise linear and convex over segments of $[0,1]$, but that, in general, discontinuities will occur at isolated points on $[0,1]$.

C. Theoretical Properties for Objective Function Parameterizations

When compared with the paucity of results for right-hand side parameterizations, the quantity of results for objective function parameterizations may come as some surprise. This is due to the constancy of the feasible region as a function of the parameterization. All of the results in this section deal with the behavior of the optimal solution and value as a function of the parameterization.

Consider the PILP:

For $\forall \, \theta \, \varepsilon \, [0,1]$ solve:

$$\min_{x \varepsilon X} (c + \theta f)x$$

(P_θ)
$$Ax \geqslant b$$
$$x_j \text{ integer}, j \, \varepsilon \, J$$

where X is a compact polytope. The following result [A. S. Manne, 1967, and Noltemeier, 1970] is a direct extension of the result for the corresponding PLP problem.

Theorem 6. For (P_θ) $\forall \, \theta \, \varepsilon \, [0,1]$ the optimal solution value $v(P_\theta)$ is piecewise linear, continuous, and concave.

Proof. Let $F \triangleq \{x \, | \, x \, \varepsilon \, X, \, Ax \geqslant b, \text{ and } x_j \text{ integer}, j \, \varepsilon \, J\}$. Replace F by its convex hull, $Co(F)$. It is well known that this can be accomplished by adding a finite number of linear constraints to the problem. The convex hull has the property that each of its extreme points corresponds to a feasible solution of F. But, for $\forall\, \theta\, \varepsilon\, [0,1]$: $\min\limits_{x\varepsilon Co(F)} (c + \theta f)x$ is just a PLP. Then since the optimal solution value of a PLP over θ is piecewise linear, continuous, and concave, it must also be such for the PILP. $\quad\|$

This result, when coupled with the next theorem, reduces the continuum of problems $(P_\theta)\, \forall\, \theta\, \varepsilon\, [0,1]$ to a finite number of problems for which optimal solutions must be found.

Theorem 7. For (P_θ), a finite set of solutions can be constructed, each member of which is optimal over some θ-range $[a,b]$, where $0 \leqslant a < b \leqslant 1$. The union of these ranges is $[0,1]$. (There may exist solutions that are optimal only at a single value of θ, but in this case there must always be an alternative optimum that remains optimal over a nondegenerate interval including this value.)

Proof. Follows from Theorem 6 and from the fact that there is a finite number of break points for $v(P_\theta)\, \forall\, \theta\, \varepsilon\, [0,1]$. See figure below. $\quad\|$

Note that in the figure each straight line corresponds to a feasible solution for (P_θ). The heavy line denotes $v(P_\theta)$.

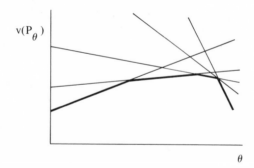

The next lemma states a well-known monotonicity property of the two portions of the objective function of (P_θ), namely, cx and fx. This property will enable us to make a statement (Theorem 8) concerning the behavior of certain variables in an optimal solution of (P_θ) as θ is varied. Let $x^*(\theta)$ be an optimal solution for (P_θ).

Lemma 1. If $0 \leqslant \theta_1 \leqslant \theta_2$, then $cx^*(\theta_1) \leqslant cx^*(\theta_2)$ and $fx^*(\theta_1) \geqslant fx^*(\theta_2)$.

Proof. Without loss of generality, take $\theta_1 = 0$ and $\theta_2 = 1$ since c and f are arbitrary. Clearly $cx^*(\theta_1) \leqslant cx^*(\theta_2)$ and $(c + \theta_2 f)x^*(\theta_2) \leqslant (c + \theta_2 f)x^*$ $(\theta_1) \leqslant cx^*(\theta_2) + \theta_2 fx^*(\theta_1)$. Therefore, $fx^*(\theta_2) \leqslant fx^*(\theta_1)$. $\quad\|$

Our next result utilizes Theorem 6 and Lemma 1 to show that it is possible to reduce the number of variables in (P_θ) for certain values of θ. This is done by fixing (pegging) variables to specific values.

Suppose (P_0) and (P_1) have been solved to optimality, and a piecewise linear, concave upper bound function $UB(\theta)$ has been found. Note that $UB(0) = v(P_0)$, $UB(1) = v(P_1)$, and that $UB(\theta)$ will have at least two "pieces," one due to $x^*(0)$ and the other due to $x^*(1)$, assuming $cx^*(0) \neq cx^*(1)$. There may be more than two "pieces" due to feasible (but not optimal) solutions actually found in the process of solving (P_0) and (P_1). See figure below.

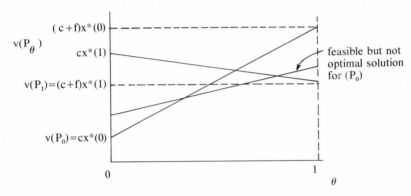

Further, suppose that x_j is a 0–1 integer variable and let $a = 0$ or 1.

Theorem 8. If $x_j^*(0) = a$, $\tilde{g}$ is any underestimate of $v(P_0 | x_j = 1 - a)$ such that $v(P_0) < \tilde{g}$, and $\tilde{h}$ is any underestimate of $v(P_1 | x_j = 1 - a)$, then x_j may be pegged to the value a for those values of θ for which $\tilde{g} + (\tilde{h} - \tilde{g})\theta \geqslant UB(\theta)$.

Proof. The linear function $\tilde{g} + (\tilde{h} - \tilde{g})\theta$ is a lower bound on $v(P_\theta | x_j = 1 - a)$ by concavity. ‖

The figure below may help in understanding this result. In this case, x_j may be pegged to the value a for $\forall\ \theta\ \varepsilon\ [0,\theta_1] \cup [\theta_2,1]$. We note that $\tilde{g}$ and $\tilde{h}$ may be found by utilizing "penalties" that are calculated during the solution process for (P_0) and (P_1).

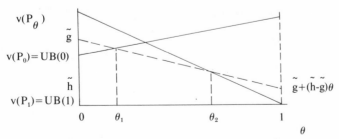

Of course, it follows that if an underestimate $\tilde{s}$ of $v(P_\theta | x_j = 1 - a)$ for some

$\hat{\theta} \; \varepsilon \; (0,1)$ is known which is greater than the value $\tilde{g} + (\bar{h} - \tilde{g})\hat{\theta}$, then the piecewise linear, concave function connecting the points $(\tilde{g},0)$, $\tilde{s},\hat{\theta})$, and $(\bar{h},1)$ may be used as an improved underestimate for $v(P_\theta | x_j = 1 - a)$ with attendant improvements in pegging x_j to the value a over segments of $[0,1]$.

The next result gives conditions under which an optimal solution of (P_θ) may be deduced to be optimal over some segment of $[0,1]$, enabling the analyst to choose judiciously the next value of $\hat{\theta}$ at which to solve (P_θ).

Let $UB(\theta)$ be a piecewise linear, concave upper bound function of $v(P_\theta)$ where $UB(0) = v(P_0)$ and $UB(1) = v(P_1)$. Let the straight lines AC and BC make up the function $UB(\theta)$ (see figure below).

Theorem 9. a) If $v(P_{\theta_C}) = v_C$, then the lines AC and BC coincide with $v(P_\theta)$ $\forall \; \theta \; \varepsilon \; [0,1]$. b) If $v(P_{\theta_C}) = v_D$, then the line AB coincides with $v(P_\theta) \; \forall \; \theta \; \varepsilon \; [0,1]$.

Proof. a) The lines AC and BC are upper bounds for $v(P_\theta)$. Since $UB(0) = v(P_0)$, $UB(1) = v(P_1)$, and $UB(\theta_C) = v(P_{\theta_C})$, and since $v(P_\theta)$ is piecewise linear, and concave, it is clear that AC and BC coincide with $v(P_\theta) \; \forall \; \theta \; \varepsilon \; [0,1]$. b) Similarly, since $UB(0) = v(P_0)$, $UB(1) = v(P_1)$, and $UB(\theta_C) = v(P_{\theta_C})$, line AB achieves the minimum for a piecewise linear, and concave function, and hence AB coincides with $v(P_\theta)$. ‖

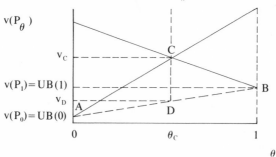

Basically, the theorem states that if $v(P_{\theta_C}) = v_C$ or $v(P_{\theta_C}) = v_D$, then (P_θ) is solved for $\forall \; \theta \; \varepsilon \; [0,1]$.

Remark. a) If $v(P_{\theta_E}) = v_E$ (see figure below), then AC coincides with the function $v(P_\theta) \; \forall \; \theta \; \varepsilon \; [0,\theta_C]$. b) If $v(P_{\theta_E}) = v_F$, then AB coincides with $v(P_\theta)$ $\forall \; \theta \; \varepsilon \; [0,1]$.

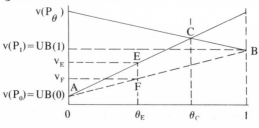

Of course, an analogous result holds for $\theta_E \,\varepsilon\, (\theta_C, 1)$.

These results for deducing optimality over a range of θ are powerful and may enable a substantial reduction to be made in computation time. They allow the analyst to solve (P_θ) at selected points only, and still be able to deduce optimality over the continuum $[0,1]$. It is clear that having solved (P_0) and (P_1), a likely choice of the next value of θ at which to solve (P_θ) is θ_C. This is because if $v(P_{\theta c}) = v_C$ or $v(P_{\theta c}) = v_D$, the entire problem for $\forall\, \theta \,\varepsilon\, [0,1]$ is solved.

Next we consider the sensitivity of an optimal ILP solution to special changes in the cost coefficients. This comes within the realm of PILP, since, by our definition, a PILP is a collection of closely related ILPs. Consider the ILPs:

$$(P) \qquad \begin{array}{l} \min\, cx \\ Ax \geqslant b \\ x_j = 0,1 \ \forall\, j\, \varepsilon\, J \end{array} \qquad\qquad (P') \qquad \begin{array}{l} \min\, c'x \\ Ax \geqslant b \\ x_j = 0,1 \ \forall\, j\, \varepsilon\, J \end{array}$$

where $c' = c$ except in the kth component. Let $\hat{x}$ be an optimal solution to (P). Piper and Zoltners [1973] state a weaker version of the following result.

Theorem 10. For some $k\, \varepsilon\, J$, suppose $\hat{x}_k = 1$. Then $\hat{x}$ is optimal in (P') if and only if $c'_k \leqslant c_k + v(P\,|\,x_k = 0) - v(P)$.

Proof. Since $\hat{x}$ is feasible in (P'), it is optimal in (P') if and only if $c'\hat{x} \leqslant v(P')$. But $v(P') = \min\, \{v(P'\,|\,x_k = 0),\, v(P'\,|\,x_k = 1)\}$ where $v(P'\,|\,x_k = 0) = v(P\,|\,x_k = 0)$ and $v(P'\,|\,x_k = 1) = c'\hat{x}$. So $\hat{x}$ is optimal in (P') if and only if $c'\hat{x} \leqslant v(P\,|\,x_k = 0)$ where $c'\hat{x} = v(P) - c_k + c'_k$. $\qquad \|$

Corollary. For some $k\, \varepsilon\, J$ suppose $\hat{x}_k = 0$. Then $\hat{x}$ is optimal in (P') if and only if $c'_k \geqslant c_k - v(P\,|\,x_k = 1) + v(P)$.

Note that in an application one may not know the value for $v(P\,|\,x_k = 0)$ in Theorem 10, but through the use of "penalties" one may calculate an overestimate of this value that can be used in the theorem. We observe that the ranges for changing the cost coefficients are valid only for one-at-a-time changes. However, by restricting this range, simultaneous changes can be made in cost coefficients. Let $(P') \triangleq (P\,|\,c'_k \leqslant c_k$ and $c'_l = c_l,\, (P'') \triangleq (P\,|\,c'_k = c_k$ and $c'_l \leqslant c_l)$, and $(P''') \triangleq (P\,|\,c'_k \leqslant c_k$ and $c'_l \leqslant c_l)$.

Theorem 11. For some $k, l\, \varepsilon\, J$ suppose $\hat{x}_k = \hat{x}_l = 1$. Then $\hat{x}$ is optimal in (P''') if $c'_k \leqslant c_k$ and $c'_l \leqslant c_l$.

Proof. By Theorem 10, $\hat{x}$ remains optimal in (P') and (P''). We will show that $v(P'''\,|\,x_k = x_l = 1) \leqslant \min\, \{v(P'''\,|\,x_k = x_l = 0),\, v(P'''\,|\,x_k = 1,\, x_l, = 0),$

$v(P'''|x_k = 0, x_l = 1)\}$. First, $v(P'''|x_k = x_l = 1) \leqslant v(P|x_k = x_l = 1) \leqslant v(P|x_k = x_l = 0) = v(P'''|x_k = x_l = 0)$. Second, $v(P'''|x_k = x_l = 1) \leqslant v(P'|x_k = x_l = 1) \leqslant v(P'|x_k = 1, x_l = 0) = v(P'''|x_k = 1, x_l = 0)$. Third, $v(P'''|x_k = x_l = 1) \leqslant v(P''|x_k = x_l = 1) \leqslant v(P''|x_k = 0, x_l = 1) = v(P'''|x_k = 0, x_l = 1)$. $\|$

Defining $(Q''') \triangleq (P|c'_k \geqslant c_k \text{ and } c'_l \geqslant c_l)$ we have:

Corollary. For some $k, l \, \varepsilon \, J$, suppose $\hat{x}_k = \hat{x}_l = 0$. Then $\hat{x}$ is optimal in (Q''') if $c'_k \geqslant c_k$ and $c'_l \geqslant c_l$.

Note the difference in allowable ranges in theorems 10 and 11. In Theorem 10, the cost coefficient of a variable may be increased by a certain amount or decreased by an arbitrary amount. In Theorem 11, however, the cost coefficients may only be decreased arbitrarily. This is a direct result of the fact that $v(P|x_k = 0)$ may vary, if say, c_l is changed as is done in Theorem 11. Since $v(P|x_k = 0)$ no longer remains constant, it cannot be used as part of the bound. Finally, we note that by combining Theorem 11 and its corollary, we may simultaneously decrease cost coefficients c_k such that $\hat{x}_k = 1$, and increase cost coefficients c_l such that $\hat{x}_l = 0$.

This completes the chapter on characteristics and properties of particular parameterizations. The utilization of the results that have been given depends almost entirely on the type of parameterization in the PILP. The next chapter details techniques for improving algorithmic efficiency that depend on the specific problem class (for example, capital budgeting, facility location, and so on) being solved.

III. Problem Dependent Techniques for Improving Algorithmic Efficiency

Given a PILP made up of ILPs belonging to a special problem class, what methods are available that would improve algorithmic efficiency? Four major techniques are problem reduction, feasibility recovery, bounding problem reoptimization, and wide range bounding.

Problem reduction refers to preliminary analysis performed on a problem that may result in variables fixed at certain values, or additional constraints or cuts added that will tighten the initial (and succeeding) relaxations. Set partitioning problems [Marsten, 1971, and Garfinkel and Nemhauser, 1969] are prime examples of instances where logical reduction is used to decrease the size of the problem. Basically, logical tests are used to conclude that certain variables must take on specific values (pegging) in an optimal solution, and that certain constraints can be eliminated. P. L. Hammer and S. Nguyen [1972] have used logical tests to generate precedence relations for general 0–1 ILPs. Examples of such relations are $x_j \leqslant x_k$ or $x_j + x_k \leqslant 1$. However, for many types of ILPs such analysis is not worth the effort expended. In other words, the ILP can be solved faster without preliminary analysis. In PILP, on the other hand, if a *single* preliminary analysis can be done for many or all of the individual problems in the PILP with little or no modification, then the analysis may become more attractive because the extra work can be amortized over the whole set of ILPs in the PILP.

Since upper bounds on the optimal solution value are used for fathoming, generating good upper bounds is of primary importance in any branch and bound algorithm. *Feasibility recovery techniques* are used to generate these upper bounds. The technique involves taking an optimal or even just a feasible solution to one problem and modifying it in such a way that it becomes feasible in another problem in the PILP. A simple example is the mixed integer linear program where the right-hand side is varied. Given the optimal solution for the original problem, fix the values of the integer variables, and then reoptimize the continuous variables using the new right-hand side. If the resulting solution is feasible, then an upper bound on the revised problem has been found.

Bounding problem reoptimization is another promising technique. Reoptimization in ILP algorithms, which utilize LP as the primary relaxation, is often used to great advantage. Generally, an advanced basis is used from the preceding candidate problem as a starting basis for the current candidate problem. Reoptimization then proceeds using the dual simplex method. In PILP, this approach would also be used, but there is yet another application. By referring to the rudimentary branch and bound algorithm in

Chapter I, we see that it is possible to choose a subset S of candidate problems from the candidate list. If these candidate problems are closely related, then a reoptimization technique would probably be an efficient method for generating an optimal solution to the relaxation for each member of the set S. Such a procedure could result in smaller storage requirements as well as reduced computation time.

The fourth technique, dubbed *wide range bounding*, is based on the formal Lagrangean dual and depends on finding feasible dual solutions that serve as valid bounds on the optimal values of the primal problems in the PILP. Dual feasible solutions are often inexpensive to calculate and, at the same time, may be surprisingly good approximations to the optimal value of the primal relaxation. Thus, these methods may be used in place of (or in conjunction with) the reoptimization techniques in the previous paragraph. The trade-off involved is that computation time is reduced at a cost of producing a weaker (dual) bound. Examples of wide range bounding applied to the continuous objective function and the continuous right-hand side parameterizations are given below.

Consider the problem:

For $\forall \, \theta \, \varepsilon \, [0,1]$ solve:

$$\min_{x \varepsilon X} (c + \theta f)x$$

(P_θ)

$$Ax \geq b$$
$$x_j \text{ integer}, j \, \varepsilon \, J$$

where X is a compact polytope.

Suppose that a set of dual multipliers $\bar{\lambda} \geq 0$ for the $Ax \geq b$ constraints have been generated for some $\hat{\theta} \, \varepsilon \, [0,1]$. We know from duality theory that $v(P_\theta) \geq v(D_\theta) \triangleq \max_{\lambda \geq 0} [\inf_{x \varepsilon X} (c + \hat{\theta} f)x + \lambda(b - Ax)] \geq \inf_{x \varepsilon X} (c + \hat{\theta} f)x + \bar{\lambda}(b - Ax)$. Also it follows for any other set of dual multipliers $\tilde{\lambda} \geq 0$ that: $v(P_\theta) \geq v(W_\theta) \triangleq \max \{ \inf_{x \varepsilon X} (c + \theta f)x + \bar{\lambda}(b - Ax); \inf_{x \varepsilon X} (c + \theta f)x + \tilde{\lambda}(b - Ax) \}$. Now since $v(\theta; \lambda) \triangleq \inf_{x \varepsilon X} (c + \theta f)x + \lambda(b - Ax)$ is a piecewise linear, concave function of θ (for fixed λ), and since $v(D_\theta)$ is also a piecewise linear, concave function of θ, we have the situation shown in the figure below.

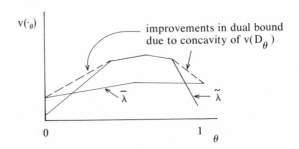

Since $v(P_\theta) \geqslant v(D_\theta) \geqslant v(W_\theta)$ and since $v(D_\theta)$ is a piecewise linear and concave, we may improve $v(W_\theta)$ by filling in the non-concave portions of $v(W_\theta)$ as shown by the dotted lines in the figure. Of course, this analysis holds for any number of choices of $\lambda \geqslant 0$ for use in the calculation of $v(W_\theta)$. In essence we are only applying the fact that $v(D_\theta)$ must be concave.

Next we consider the problem:

For $\forall\ \theta\ \varepsilon\ [0,1]$ solve:

$$\min\ cx$$
$$x\varepsilon X$$
(H_θ)
$$Ax \geqslant b + \theta r$$
$$x_j \text{ integer, } j\ \varepsilon\ J$$

where X is a compact polytope.

Suppose that we generate two sets of dual multipliers $\bar{\lambda} \geqslant 0$ and $\tilde{\lambda} \geqslant 0$ for (H_θ). Then $v(H_\theta) \geqslant v(G_\theta) \triangleq \max\ \{\bar{\lambda}(b + \theta r) + \inf_{x\varepsilon X} (c - \bar{\lambda}A)x ; \tilde{\lambda}(b + \theta r) + \inf_{x\varepsilon X} (c - \tilde{\lambda}A)x\}$. Note the interesting property that the "inf" problems do not depend on θ. This suggests that the computational burden of finding a dual bound for $(H_\theta)\ \forall\ \theta\ \varepsilon\ [0,1]$ would be cheap. Still another dual bound may be obtained from the LP tableau of $(\bar{H}_\theta)$ (if LP is used as the primary relaxation). Suppose that we have an optimal tableau for $(\bar{H}_\theta)$. The primal solution values of the basic variables are represented in the tableau by $B^{-1}(b + \hat{\theta}r)$. Now for some other value of θ, this basis may be infeasible. But if the dual simplex method is used to regain primal feasibility, we may generate a dual bound for all θ at each dual simplex iteration, since dual feasibility is retained in the dual simplex algorithm.

IV. Scheduling Solution Priorities for PILPs

In solving the PILP an important decision that must be made is the establishment of solution priorities. Specifically, should equal effort be given to solving all of the problems in the PILP at all times, or should some priority scheme be initiated whereby one problem or some subset of problems is solved to optimality, and then another subset is solved? The priority scheme to be devised should take into account the type of parameterization and its attendant characteristics and properties (Chapter II), problem class dependent techniques (Chapter III), as well as aspects identified in the first part of this chapter.

An important factor that plays an intimate role in the formulation of PILP branch and bound algorithms is whether the solution of one problem is likely to furnish information useful for solving other problems in the PILP. Three aspects of this factor will be outlined in this chapter. The first aspect is the tightness of the primary relaxation as a function of the problem set index. The second is the behavior of individual integer variables in an optimal solution as the problem set index varies. The third is the question of whether a branch and bound tree for one problem in the PILP is a "good" branch and bound tree for another problem in the PILP.

The gap between the optimal integer completion value and the optimal relaxation value at a given node in a branch and bound tree is a measure of the tightness of a relaxation. This gap, of course, is dependent on the problem set index. The behavior of the gap function is an important factor in deciding on the priority for solution of the ILPs making up the PILP, since it would seem likely that problems with smaller gaps are easier to solve than those with large gaps. This will become clearer later in the chapter when various solution priorities are outlined. In addition it may be the case that solving a "small gap" problem first may enable good feasible solutions for closely related problems to be generated by feasibility recovery techniques.

Behavior of individual integer variables in an optimal solution as the problem set index varies is also an important consideration. Intuitively we ask the following questions: (a) Do the optimal solutions remain relatively stable as the problem set index varies? (b) Do "important" variables remain "important" as the problem set index varies? Both questions can be clarified by appealing to the notion of "integer Δ's." For a 0–1 integer variable, x_j, define $\Delta(j) = v(P|x_j = 1) - v(P|x_j = 0)$. Intuitively $\Delta(j)$ may be thought of as an indicator of the "importance" of x_j in an optimal solution of (P). If $|\Delta(j)|$ is small, then the value of x_j does not have much effect on the solution value. It follows that in a branch and bound process, important variables should be the variables on which branching is done initially. Clearly, then question b would seem to be vital, for it is directly related to the third factor that will be examined now.

The question of "good" branch and bound trees for problems in a PILP can best be examined by considering the notion of "scratch tree" dynamics. Given a PILP indexed by k, and given a traditional ILP algorithm, a scratch tree is defined to be the branch and bound tree resulting from applying the ILP algorithm to, say, the kth problem in the PILP from scratch without the benefit of any prior information. Scratch tree dynamics is the study of how the scratch trees change as the problem set index, k, varies. If the scratch trees remain reasonably stable as a function of the problem set index, then it is likely that the scratch tree for the kth problem will be a "good" initial separation for the other problems. We observe that a scratch tree consists of two types of nodes — fathomed nodes and unfathomed nodes. If a node is unfathomed for one problem, it is reasonable to presume that the corresponding node will be unfathomed for a closely related problem (assuming that the same type of bounding relaxation is used). While this may not always be so, it does afford a rationale for inspecting only the fathomed nodes of the original problem. Furthermore, since the fathomed nodes form a partition of the feasible solutions of the problem, no optimal solutions can be missed by using this set of nodes as an initial candidate list (initial separation) for another problem in the PILP. Thus we may choose to use the set of fathomed nodes as an initial separation for another problem in the PILP. This is the tack used by Roodman [1972, 1974] in his study on ILP sensitivity.

Armed with a better understanding of the factors that are of major consequence in the formulation of PILP branch and bound algorithms, we now consider the scheduling of solution priorities for individual problems in a PILP. There are at least three solution priorities for a PILP branch and bound algorithm.

The first approach is *purely serial*. One problem in the PILP is solved to optimality, and then using the information gleaned from this problem the next problem is solved to optimality. This procedure continues until all problems have been solved. Information that might be of use for future problems includes: choice of an initial separation, choice of branching rules and their operating parameters, knowledge of a good upper bound, relative emphasis placed on fathoming and pegging machinery, choice of quitting threshold, and avoidance of nodes that are likely to be unfathomable.

A second approach is *lexicographically serial*. Let the problem set be indexed by k. The procedure begins as though (P_1) is the only problem to be solved. Branching, pegging, and fathoming machinery are devoted wholly to (P_1) until at some node (CP_{1,R_i}) is fathomed. Then at this node the relaxation for (CP_{2,R_i}) is solved (hopefully by reoptimizing the (CP_{1,R_i}) relaxation at the node). If (CP_{2,R_i}) is fathomed, the (CP_{3,R_i}) relaxation is solved to optimality, and so on for increasing problem indices. If (CP_{2,R_i}) is not fathomed, then branching continues from that node with the branching, pegging, and fathoming machinery devoted wholly to (P_2). When a (CP_{2,R_l}) is finally fathomed at a node, the (CP_{3,R_l}) relaxation is solved to optimality at that node. The process continues in this manner and backtracking in the branch and bound tree occurs naturally with each unfathomed node being tagged with the index of the problem to which the branching, pegging, and fathoming machinery will be initially devoted.

A purely *parallel* approach is another possibility. At each node the relaxations for (CP_{1,R_i}), (CP_{2,R_i}), ..., (CP_{K,R_i}) are solved. If some problems are fathomed, they are dropped from further consideration at any descendant of the node. When all problems that remain under consideration at a node are fathomed, backtracking from that node occurs.

The parallel approach, then, relies on the assumption that solving a series of closely related problems at a given node can be accomplished relatively efficiently. The purely serial approach on the other hand relies on the assumption that information gained from solving one member of the PILP will be helpful in solving another member of the PILP. The lexicographically serial method is one possible compromise between the two approaches.

V. The Parametric 0–1 Knapsack Problem

In this chapter we consider the parametric 0–1 knapsack problem. Three specific parameterizations are examined:

1) For $k = 1, \ldots, K$ solve:

$$(P_k) \quad \begin{array}{ll} \max & cx \\ x = 0,1 \\ & wx \leqslant B - t_k \end{array}$$

2) For $k = 1, \ldots, K$ solve:

$$(R_k) \quad \begin{array}{ll} \max & (c + f_k)x \\ x = 0,1 \\ & wx \leqslant B \end{array}$$

3) For $\forall \, \theta \, \varepsilon \, [0,1]$ solve:

$$(Q_\theta) \quad \begin{array}{ll} \max & (c + \theta f)x \\ x = 0,1 \\ & wx \leqslant B \end{array}$$

where c, f, f_k, w are comformable n-vectors and B, t_k are scalars. Without loss of generality we assume throughout this chapter that $c, w > 0$, $B > 0$, and $0 = t_1 < t_2 < \ldots < t_k < B$.

The 0–1 knapsack is a very simple model. One important application is the capital budgeting problem with one budget constraint. Problem formulations 1), 2), and 3) allow for flexibility in the budget, the cost coefficients, and in the maximizing criteria, respectively. Perhaps a more important use of the 0–1 knapsack problem is as a subproblem for a larger model. One simple example is a general capital budgeting problem (with m budget constraints). By absorbing all the budget constraints but one into the objective function, a 0–1 knapsack problem results. Under certain conditions this relaxation of the original problem can be shown to be at least as strong a relaxation as the traditional LP relaxation.

The outline for this chapter is as follows. First, an efficient algorithm for the 0–1 knapsack problem will be stated. Then algorithms for each type of parameterization will be outlined, computational results will be cited, and conclusions will be drawn concerning the most effective methods for solving each parametric 0–1 knapsack.

A. An Algorithm for the 0–1 Knapsack Problem

Consider the problem:

$$(P) \quad \begin{array}{ll} \max & cx \\ x = 0,1 \\ & wx \leqslant B. \end{array}$$

By capitalizing on the simplicity of this problem, it is possible to achieve substantial savings in computation time over, say, a general 0–1 ILP code. There are two properties of (P) that may be exploited successfully. First, we assume that the variables have been ordered by decreasing "bang-for-buck" ratios so that $(c_1/w_1) \geqslant (c_2/w_2) \geqslant \dots \geqslant (c_n/w_n)$. With this ordering the solution to the linear program $(\bar{P})$ (replacing $x = 0,1$ by $0 \leqslant x \leqslant 1$) becomes analytic. That is, variables with the largest bang-for-buck are placed in the knapsack at their upper bounds of 1 until no more room remains in the knapsack. At this point, the variable that could not fit is placed in the knapsack at a fractional level such that the knapsack is filled. It is clear that all variables, with the possible exception of one, have value 0 or 1 in an optimal solution to $(\bar{P})$. It follows that by setting the fractional variable to 0, feasibility in (P) is achieved, so that a lower bound on $v(P)$ is readily available. It is these two properties (an analytic solution to $(\bar{P})$ and a simple feasible solution generator) that are exploited in the algorithm below.

J. F. Korsh and G. P. Ingaragiola (KI) [1973] have developed an algorithm for 0–1 knapsacks that has proved to be effective in reducing computation times. Basically they employ an inexpensive LP test that, if passed, allows a variable to be pegged to 0 or 1 at the root (initial) node of a branch and bound tree. Computational results show that upwards of 80 percent of the variables may be pegged to 0 or 1. The reason for such powerful pegging is that the gap between $v(\bar{P})$ and the lower bound found by the feasible solution generator is generally small. Once the pegging tests are completed, the "reduced" knapsack problem consisting of the unpegged variables is solved by any available knapsack algorithm. Since computation time for the pegging test is linearly proportional to the number of variables, and a branch and bound approach is generally exponentially proportional, such a device would appear to be quite attractive. This is indeed true, since the KI approach reduced computation times by a factor of 5 for 50 variable problems and by over a factor of 30 for 1,000 variable problems.

R. S. Dembo [1974] has noted that the concept of Lagrangean relaxation may be used in carrying out the KI pegging tests. While his test is slightly weaker than the KI test, the computation time required for the pegging phase only is about two-thirds less than for the KI method.

The branch and bound algorithm used by KI was the H. Greenberg and R. Hegerich (GH) [1970] algorithm that, until recently, was the most efficient knapsack algorithm. However, E. Horowitz and S. Sahni (HS) [1974] have developed a branch and bound algorithm that dominates the GH algorithm. We present a variant of the HS algorithm that has decreased computation times (in the branch and bound phase) by approximately one-third over the original HS algorithm.

To make the presentation of the algorithm clear, we shall appeal to the general branch and bound framework given in Geoffrion and Marsten [1972]. An explanation of the finer points of the algorithm will be deferred until later.

Algorithm A:

1. Order the variables by decreasing bang-for-buck so that $(c_1/w_1) \geqslant (c_2/w_2) \geqslant \ldots \geqslant (c_n/w_n)$. Set $I_1 = I_0 = \phi$.

2. Solve $(\bar{P})$ getting an optimal solution $\bar{x}$ and an optimal dual multiplier $\bar{\lambda}$ associated with the budget constraint. If $\bar{x}$ is feasible in (P), stop: the solution is optimal. Otherwise denote the index of the fractional variable by r.

3. Find a lower bound z^* for $v(P)$ by setting $\bar{x}_r = 0$ in the solution to $(\bar{P})$. Let $x^* = \bar{x}$.

4. Try to improve z^* by certain heuristics.

5. For $\forall i = 1, \ldots, r - 1$, if $v(\bar{P}) - c_i + \bar{\lambda}w_i \leqslant z^*$, set $I_1 = I_1 \cup \{i\}$ (x_i is pegged to 1).

6. For $\forall i = r + 1, \ldots, n$, if $v(\bar{P}) + c_i - \bar{\lambda}w_i \leqslant z^*$, set $I_0 = I_0 \cup \{i\}$ (x_i is pegged to 0).

7. Solve the remaining knapsack problem:

$$(R) \qquad \text{Max} \quad \sum_{i\varepsilon I_1} c_i + \sum_{i\notin I_1\cup I_0} c_i x_i$$

$$\sum_{i\notin I_1\cup I_0} w_i x_i \leqslant B - \sum_{i\varepsilon I_1} w_i$$

by using the branch and bound procedure in steps 8 through 18.

8. Initialize the candidate list to consist of (R) and let the incumbent value be z^*.

9. If the candidate list is empty, stop: x^* is an optimal solution to (P) and z^* is the optimal value.

10. Select a candidate problem (CP) from the candidate list by a last-in-first-out (LIFO) rule.

11. Solve $(\overline{CP})$ getting an optimal solution $\bar{x}$.

12. If $(\overline{CP})$ is infeasible, go to 9.

13. If $v(\overline{CP}) \leqslant z^*$, go to 9.

14. If an optimal solution of $(\overline{CP})$ is feasible in (CP), go to 18.

15. Choose that x_j which is the free variable with the largest bang-for-buck.

16. If $w_j \leqslant B - \sum\limits_{\substack{i:x_i \\ \text{set to 1}}} w_i$, then add only $(CP|x_j = 0)$ to the candidate list, add the restriction $x_j = 1$ to (CP), and go to 15. Otherwise go to 17.

17. If $w_j > B - \sum\limits_{\substack{i:x_i \\ \text{set to 1}}} w_i$, add the restriction $x_j = 0$ to (CP), choose that x_j which is the free variable with the largest bang-for-buck and return to the beginning of this step. Otherwise go to 11.

18. A feasible solution to (P) has been found. Set $z^* = v(\overline{CP})$, $x^* = \bar{x}$, and go to 9.

In step 2 an optimal dual multiplier $\bar{\lambda}$ for $(\bar{P})$ can be shown to be equal to (c_r/w_r). In step 4, two heuristics are used in an attempt to improve the value of z^*. First, set $\tilde{x} = \bar{x}$ and $\tilde{x}_r = 0$. The solution $\tilde{x}$ then has a slack in the constraint with value $s = \bar{x}_r \cdot w_r$. Now, for $i = r + 1, \ldots, n$ the following is done: if $w_i \leqslant s$, set $\tilde{x}_i = 1$ and $s = s - w_i$. If $s > 0$, repeat this step for $i = i + 1$. If $c\tilde{x} > z^*$, set $z^* = c\tilde{x}$ and $x^* = \tilde{x}$. Basically, this heuristic puts extra variables in the knapsack until no more variables fit. The second heuristic begins by setting $\tilde{x} = \bar{x}$ and $\tilde{x}_r = 1$. This overfills the knapsack by $s = (1 - \bar{x}_r) \cdot w_r$. Then for $i = r - 1, r - 2, \ldots, 1$ the following is done: set $\tilde{x}_i = 0$, $s = s - w_i$, and if $s > 0$ repeat this step for $i = i - 1$. When $s \leqslant 0$, set $s = -s$ and return to the test loop in the first heuristic. Thus, this heuristic begins by overfilling the knapsack, and then variables are withdrawn until feasibility is obtained. At this point the test loop in the first heuristic is employed. The pegging tests in steps 5 and 6 utilize the notion of Lagrangean relaxation (LGR). Consider the relaxation:

$$(LGR_{\bar{\lambda}}) \qquad \max_{x = 0, 1} cx + \bar{\lambda}(B - wx).$$

It is easily seen [Geoffrion, 1974a] that $v(LGR_{\bar{\lambda}}) = v(\bar{P})$ where $\bar{\lambda}$ is an optimal dual multiplier for $(\bar{P})$, and that the solution to $(LGR_{\bar{\lambda}})$ is analytic. That is,

$$\hat{x}_i = \begin{cases} 1 \text{ if } c_i - \bar{\lambda}w_i > 0 \\ 0 \text{ if } c_i - \bar{\lambda}w_i \leqslant 0. \end{cases}$$

Thus, we have $v(LGR_{\bar{\lambda}}|x_i = 1) = v(\bar{P}) + c_i - \bar{\lambda}w_i$ if $\hat{x}_i = 0$, and $v(LGR_{\bar{\lambda}}|x_i = 0) = v(\bar{P}) - c_i + \bar{\lambda}w_i$ if $\hat{x}_i = 1$. Since $v(LGR_{\bar{\lambda}}) \leqslant v(P) \ \forall \ \lambda \geqslant 0$, it follows that the pegging tests are valid. Steps 5 and 6 may be enhanced by adding the following tests if the LGR test fails:

5b. If $v(\overline{P}\,|\,x_i = 0) \leqslant z^*$, then set $I_1 = I_1 \cup \{i\}$.

6b. If $v(\overline{P}\,|\,x_i = 1) \leqslant z^*$, then set $I_0 = I_0 \cup \{i\}$.

The branch and bound algorithm of steps 8 through 18 is straightforward, but a few comments may make it clearer. In step 10 a LIFO selection rule is used, thus guaranteeing linear storage and minimal setup costs in a computer implementation. Step 14 is an addition to the HS implementation that recognizes that if a relaxation has an optimal integer feasible solution, then the current candidate problem may be fathomed. The branching strategy in steps 15 through 17 is done as follows. The most attractive free variable (in terms of largest bang-for-buck) is chosen as the branch variable. However, since this variable has value 1 in $(\overline{CP})$, we have $v(\overline{CP}\,|\,x_j = 1) = v(\overline{CP})$. Thus, no reoptimization is required, and the next branch variable may be chosen. This continues until the next variable chosen, x_j, cannot fit in the knapsack at a level of 1. But this implies that x_j may be pegged to 0 due to feasibility considerations. Pegging variables to 0 continues until no longer possible. At this point, control is returned to step 11, and the relaxation of the current candidate problem is solved. In actuality, a group of variables (contiguous by index) are committed to 1 until this is no longer possible, and then a group of variables (contiguous by index) are pegged to 0 until this is no longer possible. This allows the LP in step 11 to be bypassed after most branching operations, which in turn reduces computation time measurably. This is seen in Table A where the HS algorithm (as coded by HS) is compared to steps 8 through 18 of Algorithm A. Note that steps 1 through 7 were omitted in these runs. Both algorithms were coded in FORTRAN H and run on an IBM 360/91. Random c_i and w_i were generated from a uniform

Table A. Comparison of HS Algorithm and Algorithm A Omitting Steps 1–7. (Time in milliseconds excluding I/O and sort time on an IBM 360/91)

Problem Identifier	Number of Variables	H&S	Algorithm A Omitting Steps 1–7
1,365	50	6	4
1,397	50	8	5
1,398	50	8	6
1,326	50	12	8
1,406	50	7	5
1,366	50	16	11
1,282	50	7	5
1,340	50	15	9
1,288	50	9	7
Total for 9 problems		88	60
Average for 9 problems		10	7

Table B. Computation Time in Milliseconds for Algorithm A
(Excluding sort and I/O time on an IBM 360/91)

Problem	Number of Variables	Percentage of Reduction in Steps 1–7	Time for Steps 2–7	Time for Steps 8–18	Total Time
1,365	50	88	2	0	2
1,397	50	82	2	1	3
1,348	50	62	2	2	4
1,326	50	56	2	4	6
1,406	50	82	2	1	3
1,366	50	58	2	6	8
1,282	50	80	3	1	4
1,340	50	82	2	3	5
1,288	50	78	2	3	5
1,468	50	86	2	1	3
Total			21	22	43
Average		75	2	2	4
2,823	100	89	3	2	5
2,772	100	65	3	2	5
2,763	100	67	3	5	8
2,945	100	88	3	1	4
2,795	100	99	2	0	2
2,706	100	85	3	5	8
2,589	100	99	3	0	3
2,992	100	95	3	1	4
2,447	100	94	3	0	3
2,771	100	86	3	1	4
Total			29	17	46
Average		87	3	2	5
5,531	200	86	4	3	7
5,790	200	79	4	10	14
5,423	200	94	4	2	6
5,641	200	99	5	0	5
5,614	200	90	4	3	7
5,536	200	94	4	2	6
5,108	200	93	4	2	6
5,274	200	95	5	1	6
5,448	200	92	4	4	8
5,734	200	93	4	4	8
Total			42	31	73
Average		92	4	3	7

distribution, $U[10,100]$, and B was set to $.5 \cdot (\sum_{i=1}^{n} w_i)$. Results show that steps 8 through 18 of Algorithm A reduced computation time by roughly 33 percent when compared with the HS algorithm.

Results for all of Algorithm A (steps 1 through 18) are given in Table B. These results clearly dominate KI's results, even when machine differences

Table C. Relative Increases in Computation Time as a Function of the Number of Variables in a Knapsack

Number of Variables	KI Algorithm	Algorithm A
50	1.0	1.0
100	1.7	1.1
200	4.2	1.7

are taken into account. Table C shows this dominance. The trend, as the number of variables increases, definitely favors Algorithm A.

Quadrupling the number of variables increases computation time by a factor of 1.7, while for KI this factor is 4.2. Finally, we mention that inclusion of steps 5b and 6b in Algorithm A was ineffective and, in fact, increased computation times slightly.

B. The 0–1 Knapsack Problem with a Finite Number of Right-Hand Sides

In this section problem 1) is considered. It is well known that optimal solutions for all right-hand sides from $1, 2, \ldots, B$ are available as a by-product if (P) is solved via dynamic programming (DP). However, with the recent developments in knapsack branch and bound technology as expounded upon in Section A, we shall see that for problems with a reasonable number of right-hand sides, the branch and bound approach is more efficient both in computation time and in storage requirements.

From a practical point of view, we note that if the various right-hand sides to be considered in parameterization 1) cover a large range, then within that range it is possible to find optimal integer solutions for certain budgets by simply filling the knapsack using the bang-for-buck ordering. These budgets correspond precisely to those right-hand sides within the range that have a naturally integer solution, so that $v(P) = v(\bar{P})$. Occasionally this might be all that is needed, and, of course, such an analysis may be done by hand calculation.

As an initial step in developing an algorithm for 1), we investigate the problem dependent techniques of Chapter III. Then factors affecting the scheduling of solution priorities of Chapter IV will be considered, three algorithms will be given, and conclusions will be drawn concerning the most efficient algorithm.

Reduction techniques can be an effective means for reducing computation time. The KI reduction used in Algorithm A is a classic example. While this test applies only for a given right-hand side, the number of computations required is minimal—1 multiplication, 1 addition, 1 subtraction, and 1 comparison per variable. Hence, even when repeated for a number of right-hand sides, the computational effort is small. Another reduction technique (which is independent of the right-hand side) is based on the following elementary result:

Theorem 12. If $c_i \geqslant c_j$ and $w_i \leqslant w_j$ then the constraint $x_i \geqslant x_j$ may be added to problem 1) without affecting the optimal solution values for $(P_k) \, \forall \, k = 1, \ldots, K$.

Proof. Assume $x_j^* = 1$ and $x_i^* = 0$ in an optimal solution for (P_k). Then let $\hat{x} = x^*$ except $\hat{x}_j = 0$ and $\hat{x}_i = 1$. Since $w_i \leqslant w_j$, this revised solution remains feasible, and since $c_i \geqslant c_j$, it has an objective value at least as great as $v(P_k)$. Since this is valid for each (P_k) individually, it must hold for $(P_k) \, \forall \, k = 1, \ldots, K$. ∥

This means that if x_i is pegged or set to 0, then x_j may be pegged to 0 also. Similarly, if x_j is pegged or set to 1, then x_i may be pegged to 1.

Feasibility recovery techniques allow one to recover a feasible solution for (P_2), say, given an optimal (or even just feasible) solution for (P_1). Suppose that a feasible solution $\hat{x}$ has been found for (P_1) where $B - t_1 \geqslant w\hat{x} > B - t_2$. One approach is to find $\{ j \, | \, \hat{x}_j = 1 \text{ and } (c_j/w_j) = \min_{i:x_i=1} (c_i/w_i) \}$. Then set

$\hat{x}_j = 0$. If $w\hat{x} \leqslant B - t_2$, $\hat{x}$ is a feasible solution for (P_2). If not, repeat the process. This approach simply removes the variable that has the worst bang-for-buck from the knapsack. Of course, a multitude of other feasibility recovery techniques can be derived as well.

Next we examine bounding problem reoptimization techniques. Due to the analytic nature of the LP solution for $(\bar{P}_k)$, it is clear that reoptimization may be handled very easily. All that is required is to keep track of the level of the fractional variable, and then to reduce its value until it reaches a level of 0, or until the budget constraint is satisfied, whichever comes first. If the budget constraint is not satisfied, the next (contiguous by index) variable equal to 1 in the original LP is reduced in the same manner.

Finally, wide range bounding methods may be employed. However, due to the efficient LP reoptimization available, such bounding techniques would not seem to be attractive for this class of problems.

We now consider factors affecting the scheduling of solution priorities. The tightness of the primary relaxation is generally a function of x_r, which is itself a function of the varying right-hand side. It is easy to see that $c_r \cdot x_r$ is an upper bound on $v(\bar{P}_k) - v(P)$, and that this upper bound varies from $c_r - \varepsilon$ to 0 (for $\varepsilon > 0$) for suitable values of the right-hand side. So the gap value fluctuates up and down (since r varies) with respect to the right-hand side, making it difficult to capitalize on the behavior of the gap.

The behavior of the individual integer variables in an optimal integer solution seems to be rather stable for the vast majority of variables. This can be seen by the consistent pegging of variables to 0 or 1 as the right-hand side varies. In general, computational experience has shown that variables with large bang-for-bucks are pegged to 1, and those with small bang-for-bucks are pegged to 0. Variables with "average" bang-for-bucks vacillate between 0 and 1 in optimal solutions as the right-hand side is varied.

The persistence of pegged variables (for varying right-hand sides) supports the contention that scratch trees are rather stable as the right-hand side varies. Furthermore, if the branch variable is always chosen to be the free variable with the largest bang-for-buck (as in Algorithm A), then since the bang-for-buck ordering remains the same, the selection of branch variables should remain stable. This line of reasoning, of course, simply reinforces the contention that scratch trees remain stable.

Three different solution approaches for problem 1) were tested, all of which used Algorithm A as a primary building block. First, a serial approach utilizing an advanced initial separation was attempted.

Algorithm B:

1. Set $k = 1$. Solve (P_k) by Algorithm A getting an optimal solution x^*.

2. If $wx^* \leq B - t_{k+1}$, x^* is optimal for (P_{k+1}) so let $k = k + 1$ and return to the beginning of this step.

3. Set $k = k + 1$. If $k > K$, stop.

4. Use a feasibility recovery technique to find a good feasible solution for (P_k). Call it x^* and the corresponding objective value, z^*.

5. Perform steps 2 through 7 of Algorithm A for (P_k).

6. Using the frontier of fathomed nodes (from (P_{k-1}) where pegged variables from steps 2 through 7 of Algorithm A are eliminated) as an initial separation (or candidate list), perform steps 9 through 18 of Algorithm A. Go to 2.

The test in step 2 is an application of Theorem 4. Step 5 employs the pegging tests of Algorithm A. In step 6 the initial candidate list used may be describes as follows (see figure on next page).

Scratch tree for (P_1) "Reduced" scratch tree

Initial Candidate List for (P_2)

$(P \mid x_1 = x_3 = x_8 = 1, x_6 = x_7 = 0)$
$(P \mid x_1 = x_3 = 1, x_6 = x_7 = x_8 = 0)$
$(P \mid x_1 = x_3 = x_7 = 1, x_6 = 0)$
$(P \mid x_1 = x_3 = x_6 = 1)$
$(P \mid x_1 = 1, x_3 = 0)$
$(P \mid x_1 = 0)$

In this figure a vertical line denotes a variable pegged to 0 or 1, and a horizontal line denotes the automatically fathomed alternate branch. The candidate list may be ordered in step 6 by the most promising relaxation value. In other words, when a candidate problem is fathomed in (P_1), say, it is stored along with its relaxation value, say $v(\overline{CP}_1)$. Then for (P_2) these candidate problems are investigated in decreasing order of relaxation value.

The reduced scratch tree is found by simply eliminating the top of the scratch tree that consists of pegged variables resulting from the pegging tests of steps 2 through 7 of Algorithm A. This is done since in general these pegged variables comprise upwards of 80 percent of the total number of variables in the problem, and, as a result, the frontier of fathomed nodes becomes rather large. By using the reduced scratch tree, the frontier is reduced significantly. Clearly the reduced frontier is a valid initial separation for (P_2). The reasoning for this approach is twofold. First, in removing the pegged variables from the scratch tree for (P_1) it is assumed that having pegged a variable in (P_1), that same variable may be pegged for (P_2). Second, in using the reduced frontier it is assumed that the unfathomed nodes in the reduced tree for (P_1) will remain unfathomed for (P_2). We should also note that occasionally "automatic" fathoming occurs for problems in the initial candidate list. For instance, consider the candidate problem $(P \mid x_1 = x_3 = x_8 = 1, x_6 = x_7 = 0)$.

Now if x_8 is pegged to 0 in (P_2), then this (CP) is fathomed by infeasibility. It is also clear that a candidate problem may be fathomed if $\sum\limits_{i:x_i \text{ set to } 1} w_i > B - t_k$.

A second solution priority is the lexicographically serial method.

Algorithm C:

1. Perform steps 2 through 7 of Algorithm A for (P_k), $k = 1, \ldots, K$ getting $(x^*)_k$ and z_k^*.

2. If an optimal solution has been found for a (P_k), remove that index k from further consideration.

3. Initialize the candidate list to consist of (R_k), $k = 1, \ldots, K$ where (R_k) is the reduced knapsack in step 7 of Algorithm A for (P_k). Individual incumbent values are denoted by z_k^*.

4. Set $k = 1$.

5. Stop if the candidate list is empty: $(x^*)_k$ is an optimal solution and z_k^* is the optimal value for $k = 1, \ldots, K$.

6. If $k > K$, set $k = 1$. Select a candidate problem (CP_k) by a LIFO rule. That is, choose the last candidate problem in the list that has a "k" subscript.

7. Solve $(\overline{CP_k})$ getting an optimal solution $\bar{x}$.

8. If $(\overline{CP_k})$ is infeasible, set $k = k + 1$ and go to 5.

9. If $v((CP_k))$, set $k = k + 1$ and go to 5.

10. If an optimal solution of (CP_k) is feasible in (CP_k), go to 14.

11. Choose that x_j which is the free variable in (CP_k) with the largest bang-for-buck.

12. If $w_j \leqslant B - t_k - \sum\limits_{i:x_i \text{ set to } 1 \text{ in } (CP_k)} w_i$, then add $(CP_h | x_j = 0)$, $h = k, \ldots, K$ to the candidate list, add the restriction $x_j = 1$ to (CP_k), $h = k, \ldots, K$, and go to 11. Otherwise go to 13.

13. If $w_j > B - t_k - \sum\limits_{i:x_i \text{ set to } 1 \text{ in } (CP_k)} w_i$, add the restriction $x_j = 0$ to (CP_h), $h = k, \ldots, K$, add $(CP_h | x_j = 1)$, $h = k + 1, \ldots, K$ to the candidate list, choose that x_j which is the free variable with the largest bang-for-buck and go to the beginning of this step. Otherwise go to 7.

14. A feasible solution to (P_k) has been found. Set $z_k^* = v(\overline{CP_k})$, $(x^*)_k = \bar{x}$, $k = k + 1$, and go to 5.

In step 2, the reduction method is applied to all K problems. Steps 3 through 14 are a modification of steps 8 through 18 of Algorithm A. Basically, the procedure is to concentrate on (P_1) in fathoming tests and branching criteria until fathoming occurs at a given node. Then at that node fathoming tests and branching criteria are applied to (P_2). This continues until all K problems have been fathomed, at which point backtracking occurs, and concentration of fathoming and branching at a node transfers back to that (P_k) with the smallest index k that has not yet been fathomed at the new node under consideration. Two advantages of such an approach are the avoidance of excessive storage of nodes, and economies in the reoptimization from (CP_k) to (CP_{k+1}). Two disadvantages are that an optimal solution for (P_k) is not always known for use in feasibility recovery techniques, and the monotonicity test for optimality cannot be applied since there is no preordained order for finding optimal solutions for (P_k), $k = 1, \ldots, K$.

A third approach is a serial method where each problem is solved from scratch. The only links between problems are the monotonicity test and the feasibility recovery technique. This approach could be thought of as the traditional approach.

Algorithm D:

1. Set $k = 1$.

2. Solve (P_k) by Algorithm A, getting x^*.

3. If $wx^* \leqslant B - t_{k+1}$, x^* is optimal for (P_{k+1}): set $k = k + 1$ and return to the beginning of this step.

4. Set $k = k + 1$. If $k > K$, stop.

5. Use a feasibility recovery technique on x^* to get a feasible solution for (P_k). Go to 2.

A fourth approach, which was not tested computationally, is the parallel method. An algorithm incorporating this method is very similar to Algorithm C. Steps 1 through 5 are identical. That is, the pegging tests for all problems are done before the branch and bound procedure begins. In the branch and bound procedure (steps 6 through 14 of Algorithm C) at any given node in the tree, $(\overline{CP_k})$ is solved for all problems that have not been fathomed at a predecessor node. All fathoming tests are applied to each of these candidate problems. A branch variable is then chosen based on one particular unfathomed candidate problem, and candidate problems for each unfathomed candidate problem are added to the candidate list. Of course, if all candidates are fathomed at a given node, branching does not occur. This approach was not tested computationally since it is very similar to Algorithm C, and since in general more candidate problem relaxations would have to be solved in this approach.

Computational results for the first three approaches are given in Table D. Ten different knapsacks of 50 variables each, and ten different knapsacks of 100 variables each were tested. Problem data was generated randomly as explained earlier. For each knapsack, 5 right-hand sides were selected by reducing each succeeding right-hand side by 1 percent for the 50 variable problems and .5 percent for the 100 variable problems. It is readily seen that Algorithm D dominates algorithms B and C by a convincing margin, and that its behavior time wise is stable. The other approaches, while occasionally approaching the times of Algorithm D, had a much larger variance in computation time. Reasons for the superiority of Algorithm D are twofold. First is the effort expended in bookkeeping and setup costs for algorithms B and C. In general, these costs outweighed the actual algorithmic calculations. A second factor is that the implementation of the LP optimization in Algorithm D is efficient, whereas the LP optimization in algorithms B and C requires more effort. In Algorithm D the LP optimization is performed only over variables with an index larger than some value. Also by construction, these variables are all "free" variables. That is, checking if a variable has been previously set to 0 or 1 is not required. In algorithms B and C, however, the LP optimization must be performed over all variables and checks must be made to ascertain whether a variable is free or not. These two factors would seem to be the major causes of the increased computation times. As further confirmation of Algorithm D's dominance, we present the following analysis. We assume that LP optimizations (or indeed any other optimizations) require the same amount of computation time at all nodes in the branch and bound tree. While this is not generally the case, it does allow for a simpler analysis. The assumption may be dropped at the cost of complicating the conclusions somewhat.

Consider an arbitrary branch and bound tree with the set of fathomed nodes denoted by N.

Lemma 2. If a branch and bound tree has $|N|$ fathomed nodes, then there are $|N| - 1$ unfathomed nodes in the tree.

Proof. By construction using the fact that $2^0 + 2^1 + \ldots + 2^{|N|-1} = 2^{|N|} - 1$. $\quad \|$

Thus, we see that the set of fathomed nodes is approximately 50 percent of the total number of nodes investigated in a branch and bound procedure. Therefore, if the serial method is used for a PILP that utilizes the set of fathomed nodes as an initial separation, then the *maximum* savings in the number of nodes investigated is 50 percent over an approach where each problem is solved from scratch. This holds for the lexicographically serial method as well. Thus, there is an upper bound on the savings that can be realized using algorithms B and C instead of Algorithm D. By coupling this bound with the additional bookkeeping and optimization costs alluded to earlier, we have some indication of why algorithms B and C failed to perform

Table D. Comparison of Algorithms B, C, D for the Knapsack Problem with Five Right-Hand Sides (Time in milliseconds excluding sort and I/O time on an IBM 360/91)

Problem	Number of Variables	Algorithm B	Algorithm B with 5b and 6b included in pegging tests	Algorithm C	Algorithm C with 5b and 6b included in pegging tests	Algorithm D	Algorithm D omitting feasibility recovery technique
1,365	50	164	42	62	51	21	19
1,397	50	36	25	27	20	14	12
1,348	50	140	178	119	96	28	26
1,326	50	98	136	55	40	26	24
1,406	50	186	186	275	150	26	24
1,366	50	133	145	111	109	29	27
1,282	50	42	95	20	28	16	15
1,340	50	137	99	54	37	20	19
1,288	50	182	NR	209	256	30	29
1,468	50	NR	NR	86	26	23	21
Total		1,118	906	1,018	813	233	216
Average		124	113	102	81	23	22
Average/right-hand side		25	23	20	16	5	4

Table D—*Continued*

Problem	Number of Variables	Algorithm B	Algorithm B with 5b and 6b included in pegging tests	Algorithm C	Algorithm C with 5b and 6b included in pegging tests	Algorithm D	Algorithm D omitting feasibility recovery technique
2,753	100	NR	NR	605	71	33	26
2,702	100	NR	NR	189	70	33	26
2,693	100	NR	NR	130	105	43	28
2,875	100	NR	NR	219	194	38	31
2,727	100	NR	NR	471	235	40	33
2,636	100	NR	NR	453	250	41	34
2,524	100	NR	NR	189	67	38	36
2,922	100	NR	NR	55	58	26	20
2,387	100	NR	NR	331	104	36	28
2,706	100	NR	NR	99	103	32	25
Total				2,741	1,257	360	287
Average				274	126	36	29
Average/right-hand side				55	25	7	6

as well as Algorithm D. We note, however, that it is entirely possible that results could prove to be different if solution methods of recent years (now obsolete) had been used as the primary building blocks for a parametric right-hand side knapsack algorithm.

It should be noted that ordering the right-hand side values in ascending or descending order is not overly important, except in the case where the change from one right-hand side to the next is very small. In such a case, there are advantages to both orderings. If the largest right-hand side problem is solved first, then the optimal solution may remain feasible for the next smaller right-hand side, and by monotonicity its explicit solution may be avoided. If the smallest right-hand side is solved first, then the optimal solution value may be an excellent lower bound for the next larger right-hand side. On balance it would appear that the potential savings would be greater if the largest right-hand side problem were solved first. It is interesting to note that the feasibility recovery technique (step 5 of Algorithm D) was not effective. Computation times were better when step 5 was deleted. The major reason for this is that the feasible solution generator for (P_k) generally gave a better solution.

Finally, we mention that a trade-off point exists for solving problem 1) by Algorithm D as opposed to DP. Horowitz and Sahni [1974] have devised a DP algorithm that effectively divides the knapsack problem into two separate problems, each being one half the size of the original problem. This allows savings not only in storage but in computation time as well. While this approach is slower than Algorithm A for (P), it does possess the property that optimal solutions for all right-hand sides $1, 2, \ldots, B$ are found as a by-product. It follows that as K increases in 1), the DP approach becomes more attractive. Using a correction ratio of 8:1 for computation times on the IBM 370/165 as compared to the IBM 360/91, we estimate (using HS's results) that the break point for a 50 variable knapsack is in the range of 12 right-hand sides. That is, if $K > 12$ for problem 1), then the DP approach becomes more attractive. However, as the number of variables increases, DP solution times increase markedly. For example, using HS's results again, the break point is 22 right-hand sides for a 60 variable knapsack.

C. The 0–1 Knapsack Problem with a Finite Number of Objective Functions

In this section we consider problem 2). In this PILP, the feasible region remains constant, while the objective function varies. Basically, algorithms B, C, and D remain the same with two exceptions. Call the modified algorithms, B', C', and D'. First, the feasibility recovery technique is modified to: for x^* optimal in (P_k), x^* is feasible in (P_{k+1}) with value $(c + f_{k+1})x^*$. Second, the variables must be reordered by descending bang-for-buck for each problem. While one may think that this reordering is unimportant computationally, we shall see that sort time is actually a substantial part of total computation time.

Algorithms B′ and C′ were not tested for this parameterization because of the poor performance of algorithms B and C for problem 1). The reasoning used in discarding these algorithms is threefold. First, the computer implementation is quite similar to that used for problem 1). Thus, since bookkeeping and setup costs took much of the computation time in algorithms B and C, it is clear that such would be the case for B′ and C′ also. Second, the feasibility recovery technique was ineffective for problem 2). This paralleled the ineffectiveness of the feasibility recovery technique for problem 1). Third, scratch tree stability is worse, generally, for problem 2) than for problem 1). This follows since in the branch and bound procedure employed, branch variables are always chosen by best bang-for-buck. Thus in problem 1), the branching remains stable since the bang-for-buck ordering remains the same. However, in problem 2) this ordering generally changes from (P_k) to (P_{k+1}). We conclude, then, that Algorithm D′ dominates B′ and C′.

Algorithm D′ was coded, and test problems were run. Problems were generated as explained earlier with the additional objective functions randomly generated from the original objective function as follows. One-third of the cost coefficients were varied by ± 10 percent, one-third by ± 5 percent, and one-third remained the same. It is interesting to note (even in the absence of comparison algorithms) that in the 100 variable problems, sort time consumed some 40 percent of total computation time (see Table E, p. 44).

Finally, we note that use of a DP algorithm for 2) would require K separate applications. Since Algorithm A dominates the best DP algorithm, Algorithm D′ would dominate it as well.

D. The 0–1 Knapsack with a Continuous Objective Function Parameterization

In this section we consider problem 3). This problem may be thought of as finding optimal solutions for all possible weightings of two criteria. For example, in a capital budgeting problem, management may be interested in maximizing some combination of net present value and of pay back over the first three years of a project.

From theorems 6 and 9 we know that $v(Q_\theta)$ is piecewise linear and convex, and that it is possible to deduce optimality over a range of θ by verifying optimality at certain values of θ. Our procedure uses Algorithm A as a major building block.

Algorithm D″:

1. Set $\theta = 0$.

2. Solve (Q_θ) by Algorithm A getting an optimal solution, $x^*(\theta)$. For each feasible solution, $\hat{x}(\theta)$, which is found, add the straight line $(c + \theta f)\hat{x}(\theta)$ to $LB(\theta)$ (the convex lower bound function for $v(\theta)$).

3. If $\theta = 0$, set $\theta = 1$ and go to 2.

Table E. Algorithm D′ for the Knapsack Problem with Ten Objective Functions

(Time in milliseconds excluding I/O but including sort on an IBM 360/91)

Problem	Number of Variables	Sort Time	Time Excluding Sort	Total Time
1,365	50	26	39	65
1,397	50	26	36	62
1,348	50	27	57	84
1,326	50	27	72	99
1,406	50	27	52	79
1,366	50	27	73	100
1,282	50	26	39	65
1,340	50	26	60	86
1,288	50	25	62	87
1,468	50	26	44	70
Total		263	534	797
Average		26	53	80
Average/objective function		3	5	8
2,823	100	61	78	139
2,772	100	64	75	139
2,763	100	60	83	143
2,945	100	61	62	123
2,795	100	60	74	134
2,706	100	62	84	146
2,589	100	61	62	123
2,992	100	61	78	139
2,447	100	62	118	180
Total		552	714	1,266
Average		61	79	141
Average/objective function		6	8	14

4. Check whether $x^*(\theta)$ can be deduced to be optimal over some range of θ. If optimal solutions have been found for $\forall\,\theta\,\varepsilon\,[0,1]$, stop. Otherwise go to 5.

5. Choose a new value of θ that is a break point of $LB(\theta)$, which has not been proven to be optimal and which is closest to the previous value of θ. Go to 2.

This algorithm relies on building up a lower bound function for $v(Q_\theta)$ by solving (Q_θ) at the end points, $\theta = 0$ and $\theta = 1$. Then, utilizing the power of Theorem 9, the next value of θ is taken to be a break point of $LB(\theta)$, which is closest to the old value of θ and which has not been proven optimal as yet.

This particular point is chosen for three reasons. First, the value of $LB(\theta)$ is likely to be a good feasible solution for (Q_θ). Second, reoptimization methods (when employed) are more efficient, in general, for a value of θ close to the previous value. Third, by choosing a break point of $LB(\theta)$ the potential for deducing optimality over a range of θ is greater.

Computational results are shown in Table F. Each component of f was randomly generated by a uniform distribution $U[10,100]$ that had a 50 percent correlation coefficient with the corresponding c_j. Note that the number of values of θ for which (Q_θ) is solved is generally slightly less than twice that of the number of optimal solutions found for $\forall\ \theta\ \varepsilon\ [0,1]$. This suggests that the method for choosing the next value of θ is quite effective. Also note that sort time for the 100 variable problems comprises about 40 percent of total computation time.

Table F. Algorithm D′ for the Knapsack Problem with a Continuous Objective Function (Time in milliseconds excluding I/O on an IBM 360/91)

Problem	Number of Variables	Total Sort Time	Time Excluding Sort	Total Time	No. of Values of θ Solved	Total No. of Optimal Solutions for (Q_θ) $\forall$ θ ε [0,1]
1,251	50	16	25	41	5	3
1,335	50	10	10	20	3	2
1,328	50	16	30	46	5	3
1,296	50	34	150	184	11	6
1,420	50	17	29	46	5	3
1,551	50	16	35	51	5	3
1,334	50	24	63	87	7	4
1,269	50	20	36	56	6	3
1,426	50	20	42	62	6	3
1,396	50	30	63	93	9	5
Total		203	483	686	62	35
Average		20	48	69	6	3.5

Table F—*Continued*

2,722	100	108	201	309	15	8
2,619	100	46	46	92	6	3
2,733	100	75	89	164	10	5
2,539	100	92	153	245	13	7
2,914	100	68	95	163	9	5
2,609	100	54	56	110	7	4
2,754	100	100	233	333	15	8
2,863	100	79	97	176	11	6
2,755	100	78	107	185	11	6
2,572	100	38	32	70	5	3
Total		738	1,109	1,847	102	55
Average		74	111	185	10	5.5

VI. The Parametric Generalized Assignment Problem

Consider the problem:

$$\min \sum_{i \varepsilon I} \sum_{j \varepsilon J} c_{ij} x_{ij}$$
$$x_{ij} = 0,1$$

(A)
$$\sum_{j \varepsilon J} r_{ij} x_{ij} \leqslant b_i \quad \forall \, i \, \varepsilon \, I$$

$$\sum_{i \varepsilon I} x_{ij} = 1 \quad \forall \, j \, \varepsilon \, J$$

where $c_{ij} \geqslant 0$, $r_{ij} > 0$, and $b_i > 0$. This problem has been referred to as the generalized assignment problem. Let the index set I denote a collection of agents, and let the index set J denote a collection of tasks. Each task j is to be assigned to exactly one agent, and each agent i may perform a collection of tasks as long as these tasks do not violate the agent's resource, b_i. The amount of resource that agent i must use to perform task j is denoted by r_{ij}, and the cost to perform the task is denoted by c_{ij}. The problem, then, is to assign each task to an agent such that agent resources are not violated, and such that the total cost of performing the tasks is minimized.

Various real-world problems can be modeled accurately as generalized assignment problems. These include scheduling variable length television commercials into time slots, assigning software development tasks to computer programmers, and scheduling payments on accounts that require "lump sum" payments. See G. T. Ross and R. M. Soland [1975] for other examples and motivations.

In this chapter, we present an efficient branch and bound algorithm [Ross and Soland, 1975] for (A) that utilizes a Lagrangean relaxation as the primary relaxation. Thus, a linear programming relaxation is not required. After this we consider three parameterizations of (A), namely:

4) For $k = 1, \ldots, K$ solve:

$$\min \sum_{i \varepsilon I} \sum_{j \varepsilon J} c_{ij} x_{ij}$$
$$x_{ij} = 0,1$$

(P_k)
$$\sum_{j \varepsilon J} r_{ij} x_{ij} \leqslant b_i - h_{ik} \quad \forall \, i \, \varepsilon \, I$$

48

$$\sum_{i \varepsilon I} x_{ij} = 1 \qquad \forall j \varepsilon J$$

5) For $k = 1, \ldots, K$ solve:

$$\min_{x_{ij} = 0,1} \sum_{i \varepsilon I} \sum_{j \varepsilon J} (c_{ij} + f_{ijk}) x_{ij}$$

(R_k)

$$\sum_{j \varepsilon J} r_{ij} x_{ij} \leqslant b_i \qquad \forall i \varepsilon I$$

$$\sum_{i \varepsilon I} x_{ij} = 1 \qquad \forall j \varepsilon J$$

6) For $\forall \theta \varepsilon [0,1]$ solve:

$$\min_{x_{ij} = 0,1} \sum_{i \varepsilon I} \sum_{j \varepsilon J} (c_{ij} + \theta f_{ij}) x_{ij}$$

(Q_θ)

$$\sum_{j \varepsilon J} r_{ij} x_{ij} \leqslant b_i \qquad \forall i \varepsilon I$$

$$\sum_{i \varepsilon I} x_{ij} = 1 \qquad \forall j \varepsilon J$$

where h_{ik}, f_{ijk}, and f_{ij} are scalars.

Algorithms are presented for problems 4), 5), and 6), computational results are given, and conclusions are drawn regarding the most efficient algorithms for each parameterization.

A. An Algorithm for the Generalized Assignment Problem

Ross and Soland [1975] present an efficient branch and bound algorithm for (A) that does not use linear programming as the primary relaxation. Rather, a relaxation that requires the solution to a number of 0–1 knapsack problems is used. We present a variant of their approach utilizing the concept of Lagrangean relaxation (Geoffrion. 1974 a).

Before stating the algorithm, the primary relaxation will be developed. Consider the relaxation of (A):

$$\min_{x_{ij} = 0,1} \sum_{i \varepsilon I} \sum_{j \varepsilon J} c_{ij} x_{ij}$$

(LGR_1)

$$\sum_{i \varepsilon I} x_{ij} = 1 \qquad \forall j \varepsilon J.$$

This relaxation simply ignores the agents' resource constraints, or equivalently a vector $\beta = 0$ is assigned to these constraints, and the term $\sum_{i \varepsilon I} \beta_i (\sum_{j \varepsilon J} r_{ij} x_{ij} - b_i)$ is placed in the objective function. Thus (LGR_1) may be solved as a special linear program by replacing $x_{ij} = 0,1$ by $0 \leqslant x_{ij} \leqslant 1$. It is obvious that an optimal solution to the linear program $(\overline{LGR_1})$ is an integer solution that satisfies the constraint $x_{ij} = 0,1$. Call this solution x^{GUB}. It is also easy to see that optimal dual multipliers, $\bar{\lambda}_j$, for $(\overline{IGR_1})$ lie anywhere in the range $[c_{i_1 j}, c_{i_2 j}]$ where $c_{i_1 j}$ is the smallest c_{ij} in column j, and $c_{i_2 j}$ is the second smallest c_{ij} in column j. A second (and generally tighter) relaxation is:

$$\min_{x_{ij} = 0,1} \left(\sum_{i \varepsilon I} \sum_{j \varepsilon J} c_{ij} x_{ij} \right) + \sum_{j \varepsilon J} \bar{\lambda}_j \left(1 - \sum_{i \varepsilon I} x_{ij} \right)$$

$$\sum_{j \varepsilon J} r_{ij} x_{ij} \leqslant b_i \quad \forall\, i \,\varepsilon\, I,$$

which is equivalent to:

$$\sum_{j \varepsilon J} \bar{\lambda}_j - \max_{x_{ij} = 0,1} \sum_{j \varepsilon J} \sum_{i \varepsilon I} (\bar{\lambda}_j - c_{ij}) x_{ij}$$

(LGR_2)

$$\sum_{j \varepsilon J} r_{ij} x_{ij} \leqslant b_i \quad \forall\, i \,\varepsilon\, I$$

where $\bar{\lambda}_j$ is chosen to be equal to $c_{i_2 j} \,\forall\, j \,\varepsilon\, J$. Note that this problem separates on I into $|I|$ independent 0–1 knapsacks. Also it is clear that all variables that are equal to 0 in x^{GUB} may be set to 0 in an optimal solution to (LGR_2). This follows since it is assumed that $r_{ij} > 0$, and since the corresponding cost coefficient $(\bar{\lambda}_j - c_{ij})$ is less than or equal to 0. Thus over all I knapsacks there are a total of only $|J|$ free variables. An optimal solution to (LGR_2) will be called x^{LGR}.

After introducing some additional notation, a branch and bound algorithm will be stated. Let S be the set of variables in a partial solution. In other words, $S = \{x_{ij} | x_{ij} \text{ is assigned a value of 0 or 1}\}$. Let $J_F = \{j | j \,\varepsilon\, J$ and $\nexists\, x_{ij} = 1$ for $x_{ij} \,\varepsilon\, S\}$. Let (A) be as given earlier and let (B) be:

$$\sum_{x_{ij} \varepsilon S} c_{ij} x_{ij} + \min_{\substack{x_{ij} = 0,1 \\ x_{ij} \notin S}} \sum_{i \varepsilon I} \sum_{j \varepsilon J_F} c_{ij} x_{ij}$$

(B)

$$\sum_{\substack{i \varepsilon I \\ x_{ij} \notin S}} x_{ij} = 1 \,\, \forall\, j \,\varepsilon\, J_F.$$

Let (C) be:

$$\sum_{x_{ij}\varepsilon S} c_{ij}x_{ij} + \sum_{j\varepsilon J_F} \bar{\lambda}_j - \max_{\substack{x_{ij}=0,1 \\ x_{ij}\notin S}} \sum_{j\varepsilon J_F} \sum_{i\varepsilon I} (\bar{\lambda}_j - c_{ij})x_{ij}$$

(C)

$$\sum_{\substack{j\varepsilon J_F \\ x_{ij}\notin S}} r_{ij}x_{ij} \leqslant b_i - \sum_{\substack{x_{lj}\varepsilon S \\ l=i}} r_{lj}x_{lj} \quad \forall\, i\,\varepsilon\, I.$$

Let $(\bar{C})$ be (C) with $x_{ij} = 0,1$ replaced by $0 \leqslant x_{ij} \leqslant 1$.

Algorithm E:

1. Initialize the candidate list to consist of (A) and let z^* be a large number. Set $S = \phi$ and $J_F = J$.

2. Stop if the candidate list is empty: if there exists an incumbent, then it is an optimal solution for (A), otherwise (A) has no feasible solution.

3. Select a candidate problem (CP) from the candidate list using a LIFO rule.

4. Solve (B) for this candidate problem and get a solution x^{GUB}. If $v(B) \geqslant z^*$ or no feasible solution is found, go to 2. If $x^{GUB} \cup (\bigcup_S x_{ij})$ is feasible in (A), go to 9. Otherwise, find a vector $\bar{\lambda}$, each component of which corresponds to a $j\,\varepsilon\, J_F$, where $\bar{\lambda}_j$ is chosen to be the second smallest c_{ij} in column j such that $x_{ij}\notin S$. If a $\bar{\lambda}_j$ does not exist for some $\hat{j}\,\varepsilon\, J_F$, set $i^* = \{i\,|\,x_{ij}^{GUB} = 1\}$ and go to 11.

5. Solve $(\bar{C})$. If $v(\bar{C}) \geqslant z^*$, go to 2.

6. Solve (C) and get a solution x^{LGR}. If $v(C) \geqslant z^*$, go to 2.

7. Try to modify x^{LGR} to x^{MOD} so that $x^{MOD} \cup (\bigcup_S x_{ij})$ is feasible in (A), and such that the corresponding objective function value is equal to $v(C)$. If successful, go to 10.

8. Separate (CP) into two new problems by finding that variable which satisfies:

$$t_{i*j*} = \max_{x_{ij}^{LGR}=1} \ (\bar{\lambda}_j - c_{ij}) \cdot \left(b_i - \sum_{\substack{x_{lj}\varepsilon S \\ l=i}} r_{lj}x_{lj}\right)\Big/r_{ij}.$$

If there is a tie, break it arbitrarily. The new problems are placed in the candidate list in the order $(CP\,|\,x_{i*j*} = 0)$ and $(CP\,|\,x_{i*j*} = 1)$. Update S and J_F for each new problem. Additional variables may be assigned to 0 in $(CP\,|\,x_{i*j*} = 1)$: a) $x_{ij*} = 0$ if $i \neq i^*$ and $x_{ij*}\notin S$; b) $x_{i*j} = 0$ if $x_{i*j}\notin S$ and $r_{i*j} > b_{i*} - \sum_{x_{i*j}\varepsilon S} r_{i*j}x_{i*j}$ for $j \neq j^*$. Update S, and go to 2.

9. Set $v(C) = v(B)$ and $x^{MOD} = x^{GUB}$.

10. An improved feasible solution for (A) has been found. Record this solution as the new incumbent, $x^* = x^{MOD} \cup (\bigcup_S x_{ij})$ and set $z^* = v(C)$. Go to 2.

11. Add the problem $(CP|x_{i*j} = 1)$ to the candidate list. Update S and J_F. Additional variables may be set to 0: $x_{i*j} = 0$ if $x_{i*j} \notin S$ and $r_{i*j} > b_{i*} - \sum_{x_{i*j}\varepsilon S} r_{i*j}x_{i*j}$ for $j \neq \hat{j}$. Update S, and go to 2.

In Step 4, (B) is solved and the usual fathoming tests are employed. If all fathoming tests fail, an attempt is made to tighten the relaxation by solving $(\bar{C})$. However, if some $\bar{\lambda}_j, \hat{j} \varepsilon J_F$ does not exist, this implies that only one agent i^* is available to handle task $\hat{j}$. Hence, x_{i*j} may be pegged to 1 in step 11. As a result of this peg to 1, additional variables may be pegged to 0, thus hopefully tightening the relaxation even further. If a peg to 1 is not possible, and all fathoming tests for $(\bar{C})$ fail, then (C) is solved in step 6. Note that Algorithm A is used to solve each individual knapsack. Of course, the knapsack algorithm need not be used for agent i if the x^{GUB} solution does not violate that agent's resource. In step 7, tasks that are not assigned in the solution to (C) (this occurs since the $\sum_{i \varepsilon I} x_{ij} = 1 \ \forall \ j \varepsilon J$ constraints have been relaxed) are reassigned to the available agent with the second smallest cost. If all such tasks are reassigned and the knapsack constraints are satisfied, then this modified solution is feasible and has an objective value precisely equal to $v(C)$. Thus a new incumbent is generated. In step 8 when all fathoming tests fail, branching must occur. The branching criterion uses a weighted penalty for not choosing an agent i^* to perform a task j^* multiplied by the current amount of resource available to agent i^*. Note also that the ordering of the new candidate problems in the candidate list is such that the $x_{i*j*} = 1$ branch is investigated first.

Computational results are presented in Table G. The 50 variable problem is taken from Ross and Soland [1975]. The nine versions differ only in the agent resources as shown. All other test problems were generated per Ross and Soland. The r_{ij}, c_{ij} were randomly generated with distributions $U[5,25]$ and $U[10,50]$, respectively. Each b_i was set equal to $9*|J|/|I| + .4*\{\max_{i\varepsilon I} \sum_{j \varepsilon J} r_{ij}\hat{x}_{ij}\}$ where $\hat{x}$ is an optimal solution to (LGR_1). The results in Table G show that the algorithm is very efficient for most problems. Note that our algorithm improves upon the Ross and Soland version for the 50 variable problems by investigating fewer nodes and solving fewer 0–1 knapsacks. Coupled with Ross and Soland's results, we conclude that the algorithm is an order of magnitude faster than other algorithms such as RIP30C, which is a general purpose 0–1 code, and IPNETG, which is a branch and bound generalized network code.

B. The Generalized Assignment Problem with a Finite Number of Right-Hand Sides

In this section we consider problem 4). By allowing the right-hand side to vary, flexibility is introduced into the model. This allows the analyst to examine the effect of changes in agent resources. We begin a formal analysis of problem 4) by investigating the problem dependent techniques of Chapter III. Then, the factors affecting the scheduling of solution priorities will be discussed, and three algorithms will be presented. Finally, computational experience will be cited, and conclusions drawn concerning the most efficient algorithm.

The use of reduction techniques in problem 4) does not appear to be promising. We mention one such technique and then point out its shortcomings. Consider agent i and tasks j_1 and j_2. Let $M_i \triangleq \max_{k=1,\ldots,K} \{b_i - h_{ik}\}$. The value M_i is the maximum resource available to agent i over all K problems. If $r_{ij_1} + r_{ij_2} > M_i$, then $x_{ij_1} + x_{ij_2} \leqslant 1$ is a valid constraint for $(P_k) \; \forall \, k = 1, \ldots, K$. However, the addition of such a constraint destroys the structure of the problem, and consequently the solutions to various relaxations (for example, (LGR_1)) are no longer analytic in nature. The crux of the situation is that in attempting to tighten the primary relaxation, the structure of the problem is destroyed, and hence solution efficiencies are lost. Thus, certain reduction techniques can actually complicate the problem, and, in fact, their incorporation into an algorithm may retard efficiency.

We now consider feasibility recovery techniques. Suppose that we have found a feasible solution $\hat{x}$ for (P_1) that is not feasible in (P_2). Further, suppose that not all agent resources are violated by $\hat{x}$ in (P_2). One approach is to reassign tasks from an overassigned (infeasible) agent one at a time such that each reassignment is made at the smallest additional cost, and such that it is made to an agent with sufficient resource available. Other approaches can also be devised ad infinitum.

Next are bounding problem reoptimization techniques. Suppose that at a generic node in a branch and bound tree, (LGR_2) has been solved to optimality for a candidate problem associated with (P_1). Can this solution be reoptimized efficiently for the candidate problem at that node which corresponds to (P_2)? In general the answer is no. There are two reasons for this. First, additional pegging of variables to 0 by infeasibility (as in step 8b of Algorithm E) may occur if resources are reduced from (P_1) to (P_2). Conversely, if resources are increased, variables may no longer be pegged to 0. Of course, if feasibility pegs are not enforced, this reasoning does not apply. Second, as we have seen in Chapter V, solving closely related 0–1 knapsacks is best accomplished by solving them separately. We observe, however, that if the change in the feasible region is monotone, and x^{LGR} remains feasible in (P_2),

Table G. Computational Results for Algorithm E
(Time in seconds excluding I/O)

Problem	Agents	Tasks	Variables	Ross and Soland Algorithm (CDC 6600)			Algorithm E (IBM 360/91)		
				Number of 0–1 Knapsacks	Number of Nodes	Total Time	Number of 0–1 Knapsacks	Number of Nodes	Total Time
1. 50 variable problem from Ross and Soland [1975]									
112	5	10	50	1	1	.003	1	1	.003
104	5	10	50	7	7	.012	7	7	.015
96	5	10	50	9	9	.016	9	9	.018
88	5	10	50	5	5	.015	5	5	.009
80	5	10	50	7	5	.012	7	5	.011
72	5	10	50	41	37	.059	31	29	.044
64	5	10	50	34	29	.049	31	27	.042
56	5	10	50	36	29	.051	31	29	.045
48	5	10	50	1,084	725	.333	307	690	.297

Table G—*Continued*

Problem	Agents	Tasks	Variables	Ross and Soland Algorithm (CDC 6600)			Algorithm E (IBM 360/91)		
				Number of 0-1 Knapsacks	Number of Nodes	Total Time	Number of 0-1 Knapsacks	Number of Nodes	Total Time
2. Randomly generated problems									
171	5	50	250				2	1	.014
170	5	50	250				61	51	.170
176	5	50	250				1	1	.013
172	5	50	250				56	45	.173
165	5	50	250				2	1	.014
181	5	50	250				72	49	.182
161	5	50	250				39	57	.163
97	10	50	500				2	1	.027
82	10	50	500				—	—	>9.0
97-2	10	50	500				51	33	.142
85	10	50	500				6	11	.050
90	10	50	500				465	305	1.170
84	10	50	500				514	187	1.021
102	10	50	500				20	19	.080

then x^{LGR} is optimal for the candidate problem corresponding to (P_2).

Wide range bounding techniques may also be employed. Once again suppose that a candidate problem relaxation associated with (P_1) has been optimized at a generic node. We wish to calculate a dual bound for the candidate problem associated with (P_2) at that node. Such a bound may be calculated (compare Chapter III), but one must be careful to use "good" dual multipliers in calculating the bound. (Dual multipliers are said to be "good" if the corresponding Lagrangean relaxation, which uses these multipliers, generates a tight bound.) If (LGR_2) has been solved to optimality, then "good" dual multipliers are not readily available since (LGR_2) consists of $|I|$ independent 0–1 knapsacks. It is generally true that deriving "good" dual multipliers for an ILP is a difficult task. Nonetheless, if $(\overline{LGR_2})$ has been solved, dual multipliers for the continuous relaxation are available, and a dual bound may be calculated for the candidate problem corresponding to (P_2).

We now turn to the factors affecting the scheduling of solution priorities. Tightness of the primary relaxation as a function of the right-hand side can best be explained by using the formulation:

$$\min_{x_{ij}\,=\,0,1} \sum_{i \in I} \sum_{j \in J} c_{ij} x_{ij}$$

(S_b)
$$\sum_{j \in J} r_{ij} x_{ij} \leqslant b \quad \forall \, i \, \varepsilon \, I$$

$$\sum_{i \in I} x_{ij} = 1 \quad \forall \, j \, \varepsilon \, J$$

where b is a nonnegative scalar. As b increases, the solution x^{LGR} for (LGR_2^b) comes closer to feasibility because the knapsack constraints become less constraining. In fact, for a sufficiently large value of b, x^{LGR} is feasible in (S_b) and hence is optimal in (S_b). The relationship between $v(S_b)$ and $v(LGR_2^b)$ is indicated in the figure below.

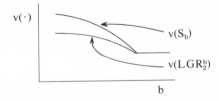

(Note that the graph will not be continuous, in general, but is drawn that way for the sake of simplicity.) By referring to Table G (p. 54) for the 50 variable problems, we see that as the agent resources decrease, the number of nodes investigated generally increases. This empirical behavior, when coupled with the graph above, substantiates the belief that as the tightness of the primary relaxation deteriorates, a problem becomes more difficult to solve (in terms of the number of nodes investigated).

The behavior of the individual integer variables in an optimal solution is difficult to categorize for this particular parameterization. As resources are reduced, some tasks may be reassigned to other agents. However, it is not a simple matter to deduce which tasks are reassigned.

The persistence of scratch trees as b varies seems to be good, empirically speaking, for small changes in the right-hand side. This is because the solution to (LGR_1^b) remains rather stable, and in the solution to (LGR_2^b) the "most attractive" variables remain at a level of 1. Since the branching rule is to branch on the "most attractive" variable that is equal to 1 in x^{LGR}, the ordering of potential branch variables remains stable, and hence the scratch trees remain stable. The cause of instability in general is the pegging of variables to 0 by infeasibility. This may cause x^{GUB} to change, which in turn changes x^{LGR}. We note that if the pegging by infeasibility were not implemented, the scratch trees would tend to be more stable. Of course this stability is bought for the price of larger scratch trees, since fathoming power at a node is reduced when the logical pegs to 0 are not enforced.

Next, we present three approaches for solving problem 4), each of which uses Algorithm E as a primary building block. The first is a serial approach that utilizes an initial separation consisting of the fathomed nodes from the previous (P_k). Recall that we define (P_k) and (P_{k+1}) to be relatively monotone if $F(P_k) \supseteq F(P_{k+1})$. This approach is similar in spirit to Algorithm B of Chapter V, Section B.

Algorithm F:

1. Set $k = 1$. Solve (P_k) by Algorithm E, getting an optimal solution $(x^*)_k$.

2. If $k > K$, stop. If (P_k) and (P_{k+1}) are relatively monotone and if $(x^*)_k$ $\varepsilon F(P_{k+1})$, then $(x^*)_k$ is optimal in (P_{k+1}), so let $k = k + 1$ and return to the beginning of this step.

3. Set $k = k + 1$. If $k > K$, stop.

4. Use a feasibility recovery technique on $(x^*)_{k-1}$ to find a good feasible solution for (P_k). Call it $(x^*)_k$ and its corresponding objective value, z_k^*.

5. Use the frontier of fathomed nodes from (P_{k-1}) as an initial candidate list and order the list in increasing order of optimal relaxation value calculated for the (CP_{k-1}) problems. With this ordered candidate list, solve (P_k), using steps 2 through 11 of Algorithm E. Go to 2.

Note that under monotonicity, step 2 may allow one to avoid explicit solution of a particular (P_{k+1}). Step 4 can be expanded to allow a feasibility recovery technique for (P_{k+1}) to be invoked whenever a feasible solution for (P_k) is found. This option is incorporated in our computer implementation. In step 5, if (P_k) and (P_{k+1}) are relatively monotone, it may be possible to "automatically" fathom a candidate problem if $v((CP_{k-1})) \geqslant z_k^*$. This follows since $v(P_k) \geqslant v(P_{k-1})$. Also, of course, an automatic fathom occurs if $\sum_{x_{ij} \varepsilon S} r_{ij} x_{ij} > b_i - h_{ik}$ for some $i \, \varepsilon \, I$.

A second approach is lexicographically serial, which is similar in spirit to Algorithm C in Chapter V, Section B.

Algorithm G:

1. Set $k = 1$.

2. Initialize the candidate list to consist of $(P_k) \; \forall \, k = 1, \ldots, K$. Individual incumbent solutions are denoted by $(x^*)_k$ and solution values by z_k^*. Set $J_F = \phi, S_k = \phi \; \forall \, k = 1, \ldots, K$. Go to 4.

3. Set $k = k + 1$. If $k > K$, set $k = 1$.

4. Stop if the candidate list is empty: $(x^*)_k$ is an optimal solution, and z_k^* is the optimal value (if they exist, otherwise (P_k) is infeasible) for $\forall \, k = 1, \ldots, K$.

5. If $k > K$, set $k = 1$. Select a candidate problem (CP_k) by a LIFO rule. That is, choose the last candidate problem in the list that has a k subscript. If none exists, set $k = k + 1$ and return to the beginning of this step.

6. Solve $(B)_k$ for this candidate problem and get a solution x^{GUB}. If $v((B)_k) \geqslant z_k^*$ or no feasible solution is found, go to 3. If $x^{GUB} \cup (\bigcup_S x_{ij})$ is feasible in (P_k), go to 12. Otherwise, find a vector $\bar{\lambda}$, each component of which corresponds to a $j \, \varepsilon \, J_F$, and where $\bar{\lambda}_j$ is chosen to be the second smallest c_{ij} in column j, such that $x_{ij} \notin S_k$. If a $\bar{\lambda}_j$ does not exist for some $\hat{j} \, \varepsilon \, J_F$, set $i^* = \{i \, | \, x_{ij}^{GUB} = 1\}$, and go to 13.

7. Solve $(\bar{C})_k$. If $v((\bar{C})_k) \geqslant z_k^*$, go to 3.

8. Solve (C) and get a solution x^{LGR}. If $v((C)_k) \geqslant z_k^*$, go to 3.

9. Try to modify x^{LGR} to x^{MOD} so that $x^{MOD} \cup (\bigcup_{S_k} x_{ij})$ is feasible in (P_k), and such that the corresponding objective value is equal to $v((C)_k)$. If successful, go to 12.

10. Select a branch variable by finding:

$$t_{i*j*} = \max_{x_{ij}^{LGR} = 1} (\bar{\lambda}_j - c_{ij}) \cdot (b_i - h_{ik} - \sum_{\substack{x_{lj} \varepsilon S \\ l = i}} r_{lj} x_{lj}) / r_{ij}$$

If there is a tie, break it arbitrarily. Add the problems $(CP_h | x_{i*j*} = 0) \; \forall \, h = k, \ldots, K$ and $(CP_h | x_{i*j*} = 1) \; \forall \, h = k, \ldots, K$ to the candidate list.

Update J_F and S_h $\forall h = k, \ldots, K$. Additional variables may be pegged to 0 for individual (P_k) as in step 8 of Algorithm E. Update S_h $\forall h = k, \ldots, K$ and go to 4.

11. Set $v((C)_k) = v((B)_k)$ and $x^{MOD} = x^{GUB}$.

12. An improved feasible solution to (P_k) has been found. Record this solution as the new incumbent, $(x^*)_k = x^{MOD} \cup (\bigcup_{S_k} x_{ij})$, and set $z^* = v((C)_k)$. If $k < K$, use a feasibility recovery technique to find a feasible solution for (P_{k+1}). If this solution improves the incumbent, set it equal to $(x^*)_{k+1}$ and its objective value to z^*_{k+1}. Go to 3.

13. Add the problems $(CP_h|x_{i*j} = 0)$ $\forall h = k + 1, \ldots, K$ and $(CP_h|x_{i*j} = 1)$ $\forall h = k, \ldots, K$. Update J_F and S_h $\forall h = k, \ldots, K$. Additional variables may be assigned to 0 for individual (P_k) as in step 11 of Algorithm E. Update S_h $\forall h = k, \ldots, K$ and go to 4.

Basically, the procedure is to concentrate on (P_1) in fathoming tests and branching criteria until fathoming for (CP_1) occurs at a given node. Then at that node fathoming tests and branching criteria are applied to the corresponding (CP_2). Note that if (P_1) and (P_2) are relatively monotone, and if $v(CP_1) \geqslant z^*_2$, then (CP_2) is automatically fathomed. Automatic fathoming also occurs if no feasible solution to a relaxation can be found. In step 13 x_{i*j} is pegged to 1 for (CP_k). However, the peg may not be valid for (CP_h), $h = k + 1, \ldots, K$. Thus, $(CP_h|x_{i*j} = 0)$ and $(CP_h|x_{i*j} = 1)$ $\forall h = k + 1$, $\ldots, K$ are added to the candidate list. It is clear, however, that if relative monotonicity holds between, say, (P_k) and (P_h), $h > k$, then $(CP_h|x_{i*j} = 0)$ need not be added to the candidate list.

A third approach for 4) is the serial method where each problem is solved from scratch. This approach is similar to Algorithm D in Chapter V, Section B, and it may be dubbed the traditional approach.

Algorithm H:

1. Set $k = 1$.

2. Solve (P_k) by Algorithm E getting $(x^*)_k$.

3. If $k > K$, stop. If (P_k) and (P_{k+1}) are relatively monotone, and if $(x^*)_k \varepsilon F(P_{k+1})$, then $(x^*)_k$ is optimal in (P_{k+1}), so let $k = k + 1$, and return to the beginning of this step.

4. Set $k = k + 1$. If $k > K$, stop.

5. Use a feasibility recovery technique on $(x^*)_{k-1}$ to get a good feasible solution for (P_k). Call it $(x^*)_k$, and the corresponding objective value, z^*_k. Go to 2.

Alternatively, step 5 may be placed in step 10 of Algorithm E so that the feasibility recovery technique is invoked for (P_{k+1}) whenever a feasible solution for (P_k) is found. This option is used in our computer implementation.

Computational results for algorithms F, G, and H are given in Table H. Problems were generated randomly as explained in section A. The first right-hand side (for (P_1)) was generated as explained. Succeeding right-hand sides were obtained by reducing all resources by approximately 2.5 percent for the 250 variable problems and 5 percent for the 500 variable problems. Table I (p. 63) gives the ratios of the relative times for each algorithm. Note that Algorithm F generally dominates algorithms G and H, except for the 50 variable problems where Algorithm G was best. This general domination by Algorithm F becomes more pronounced as the number of variables increases. This behavior is mainly due to three factors. First is the effect of bookkeeping and setup costs. As problem size increases, computation time for the primary relaxations (LGR_1) and (LGR_2) increases, while bookkeeping and setup costs remain about the same. Second, the monotonicity of the feasible regions for this parameterization allows the automatic fathoming tests to be used. If this test is successful, the primary relaxations need not be solved. Third, tree stability tends to be good for this parameterization due to the stability of the x^{GUB} and x^{LGR} solutions as the right-hand side varies.

C. The Generalized Assignment Problem with a Finite Number of Objective Functions

In this section we consider problem 5). Basically, algorithms F, G, and H remain the same with three exceptions. Call the modified algorithms F′, G′, and H′. First, the feasibility recovery technique is modified so that it simply costs out a feasible solution to (R_k) using the objective function for (R_{k+1}). Second, the monotonicity test is no longer pertinent, and third the automatic fathoming test can no longer be used. However, note that the pegs to 0 and 1 remain valid for $\forall k = 1, \ldots, K$, since these pegs are based only on feasibility considerations.

We remark that reduction techniques as examined in section B do not seem to be promising for much the same reasons. In addition, bounding problem reoptimization and wide range bounding techniques would seem to be inefficient as well, due to many of the reasons given in Section B.

Analysis of the tightness of the primary relaxation in Section B confirms that the tightness is a function of how constraining the agent resources are. Since in problem 5) these resources remain constant, the differences in tightness would seem to lie in the initial x^{GUB} solution for each (R_k), $k = 1, \ldots, K$. If a particular x^{GUB} solution "almost" satisfies the knapsack constraints, then one might surmise that the initial relaxation is tight. Alternatively, if an x^{GUB} solution heavily violates the knapsack constraints, then the relaxation will be loose. Behavior of the individual integer variables in an

Table H. Comparison of Algorithms F, G, H for the Generalized Assignment Problem with Five Right-Hand Sides: Monotone Feasible Regions

(Time in seconds excluding I/O on an IBM 360/91)

Problem	Agent Resources	Algorithm F		Algorithm G[+]		Algorithm H	
		Number of Nodes	Total Time	Number of Nodes	Total Time	Number of Nodes	Total Time
A. 50 variable problem from Ross and Soland [1975]							
1.	88	5	.009			5	.008
	80	2	.011			5	.010
	72	41	.077			29	.044
	64	7	.017			27	.042
	56	19	.031			29	.045
	Setup[#]		.011				.007
	Totals	74	.156	78	.110	95	.156
B. Randomly generated 250 variable problems with 5 agents and 50 tasks							
2.	185	3	.018			3	.015
	180	0*	0			0*	0
	175	55	.189			57	.200
	170	0*	0			0*	0
	165	45	.141			51	.175
	Setup		.023				.023
	Totals	103	.371	146	.387	111	.413
3.	185	21	.064			21	.065
	180	0*	0			0*	0
	176	3	.009			41	.155
	172	60	.240			45	.160
	167	0*	0			0*	0
	Setup		.022				.023
	Totals	84	.335	228	.527	107	.403
4.	179	1	.007			1	.004
	174	25	.070			25	.073
	170	0*	0			0*	0
	165	0*	0			1	.006
	160	0*	0			1	.006
	Setup		.021				.019
	Totals	26	.098	65	.135	28	.108

[+] Due to the nature of the lexicographic serial approach, computation times for individual problems cannot be broken out.

* Optimal solution from previous problem is feasible, hence optimal.

Feasibility generator time plus setup time.

Table H—*Continued*

Problem	Agent Resources	Algorithm F		Algorithm G[+]		Algorithm H	
		Number of Nodes	Total Time	Number of Nodes	Total Time	Number of Nodes	Total Time
5.	195	1	.007			1	.005
	189	1	.006			1	.005
	185	1	.007			1	.007
	181	55	.224			55	.210
	176	38	.128			43	.168
	Setup		.021				.020
	Totals	96	.393	83	.304	101	.415
6.	175	41	.107			41	.109
	170	0*	0			0*	0
	166	31	.101			43	.111
	161	35	.103			57	.160
	157	221	.787			253	.920
	Setup		.030				.032
	Totals	328	1.128	459	1.391	394	1.332

C. Randomly generated 500 variable problems with 10 agents and 50 tasks

Problem	Agent Resources	Algorithm F		Algorithm G[+]		Algorithm H	
		Number of Nodes	Total Time	Number of Nodes	Total Time	Number of Nodes	Total Time
7.	102	1	.009			1	.006
	97	1	.007			1	.006
	92	1	.011			1	.006
	87	21	.104			21	.080
	82	25	.080			75	.270
	Setup		.042				.042
	Totals	49	.253	100	.365	99	.410
8.	107	1	.008			1	.006
	102	1	.007			1	.006
	97	33	.146			33	.125
	92	16	.078			51	.206
	87	39	.242			63	.387
	Setup		.047				.047
	Totals	90	.528	129	.586	149	.777
9.	95	11	.031			11	.028
	90	0*	0			0*	0
	85	1	.008			11	.028
	80	23	.067			33	.080
	75	1	.005			17	.060
	Setup		.045				.044
	Totals	36	.156	78	.221	72	.240

Table H—*Continued*

Problem	Agent Resources	Algorithm F		Algorithm G$^+$		Algorithm H	
		Number of Nodes	Total Time	Number of Nodes	Total Time	Number of Nodes	Total Time
10.	112	1	.007			1	.006
	107	17	.057			17	.050
	102	5	.021			19	.081
	97	35	.142			47	.200
	92	8	.026			47	.198
	Setup		.040				.040
	Totals	66	.293	110	.417	131	.575

Table I. Ratios of Computation Times for Algorithms F, G, H with H Scaled to 1

Problem	Number of Variables	Algorithm F	Algorithm G	Algorithm H
1.	50	1.00	.67	1.
2.	250	.90	.94	1.
3.	250	.83	1.31	1.
4.	250	.92	1.25	1.
5.	250	.97	.73	1.
6.	250	.85	1.04	1.
7.	500	.61	.89	1.
8.	500	.68	.75	1.
9.	500	.65	.92	1.
10.	500	.56	.73	1.

optimal solution seems to be as obscure as for problem 4). The stability of scratch trees deteriorates for problem 5) as opposed to the stability for problem 4). We propose the following reasons for this behavior. Branch variables are chosen from among the set of variables for which $x_{ij}^{LGR} = 1$. Now $x_{ij}^{LGR} = 1$ (in (LGR_2)) implies that $x_{ij}^{GUB} = 1$ (in (LGR_1)). However, when the objective function is varied from (R_1) to (R_2), say, x^{GUB} generally changes also. This in turn implies that x^{LGR} varies. Further, x^{LGR} may vary due to the values $\bar{\lambda}_j$ that are generated directly from the objective function.

We now present a sufficient condition for an optimal solution to remain optimal for a certain parameterization. Suppose an optimal solution, x^*, for

(A) has been found. Consider the problem (A') that for some task $j_1 \, \varepsilon \, J$ has the costs $c'_{ij_1} = c_{ij_1} + d_i \; \forall \, i \, \varepsilon \, I$ where $d_i \geqslant 0 \; \forall \, i \, \varepsilon \, I$. Thus the cost for agent i_1 to perform task j_1 is increased by d_{i_1} units. Such an occurrence could be due to increased processing cost for task j_1. We have the following result.

Theorem 13. Suppose $x^*_{i_1 j_1} = 1$. If $v(A) + d_{i_1} \leqslant v(A \,|\, x_{i_1 j_1} = 0) + \min_{i \neq i_1} d_i$, then x^* is optimal in (A').

Proof. Since $\sum_{i \varepsilon I} x_{ij_1} = 1$ in any optimal solution, and since $v(A \,|\, x_{i_1 j_1} = 0) \geqslant v(A)$, any optimal solution to $(A' \,|\, x_{i_1 j_1} = 0)$ must have a cost of at least $v(A \,|\, x_{i_1 j_1} = 0) + \min_{i \neq i_1} d_i$. But x^* remains feasible in (A') with a cost of $v(A) + d_{i_1}$. Hence if the hypothesis holds, x^* is optimal in (A'). $\quad \|$

Note that any underestimate of $v(A \,|\, x_{i_1 j_1} = 0)$, such as the value $v(A)$, may be used in the theorem hypothesis.

Two problem sets were generated per Ross and Soland [1975]. For the first set, additional objective functions for each problem were generated randomly using the original objective function. One-third of the cost coefficients were varied by ± 10 percent, one-third by ± 5 percent, and one-third remained the same. Table J (p. 65) gives individual results for problems 11 through 18 and Table K (p. 68) gives ratios of computation times for the algorithms. We see that generally Algorithm H' performs the best, with algorithms F' and G' sometimes requiring more than twice as much computation time. This was found to be attributable to tree instability. That is, the individual scratch trees were not stable from one problem to the next. This points out the importance of scratch tree stability in formulating parametric branch and bound algorithms. Basically, the evidence gleaned from these test problems is that the set of fathomed nodes from (P_k) is not a "good" initial separation for (P_{k+1}). We conclude that Algorithm H' would seem to be preferable for this particular parameterization.

In the second problem set, an additional objective function was generated by increasing all costs for one agent by 10 percent. Table J gives individual results for problems 19 through 23, and Table K gives ratios of computation times. For this set of problems, Algorithm G' dominates Algorithms F' and H'. This contrasts with problems 11 through 18 where Algorithm H' generally was best. The reason for the difference is greater tree stability, which is a result of the less "radical" parameterization in problems 19 through 23. Since costs are raised by 10 percent for only one agent, the x^{GUB} and x^{LGR} solutions remain stable at a given node, and hence tree stability is more pronounced.

In conclusion, Algorithm H' seems to be best for more "radical"

Table J. Comparison of Algorithms F', G', H' for the Generalized Assignment Problem with a Finite Number of Right-Hand Sides

(Time in seconds excluding I/O on an IBM 360/91)

Problem	Objective Function Identifier	Algorithm F'		Algorithm G'[+]		Algorithm H'	
		Number of Nodes	Total Time	Number of Nodes	Total Time	Number of Nodes	Total Time

A. Randomly generated 250 variable problems with 5 agents and 50 tasks with $K = 5$

Problem	Objective Function Identifier	Number of Nodes	Total Time	Number of Nodes	Total Time	Number of Nodes	Total Time
11.	a	1	.008			1	.007
	b	1	.012			1	.007
	c	1	.012			1	.007
	d	1	.015			1	.007
	e	1	.015			1	.007
	Setup [#]		.022				.018
	Totals	5	.084	5	.056	5	.053
12.	a	43	.115			43	.080
	b	22	.068			1	.007
	c	22	.056			1	.007
	d	22	.052			1	.007
	e	42	.095			21	.040
	Setup		.030				.028
	Totals	151	.416	178	.384	67	.169
13.	a	455	1.630			455	1.192
	b	377	1.372			377	.905
	c	881	3.056			665	1.600
	d	439	1.376			213	.511
	e	754	2.888			105	.252
	Setup		.030				.030
	Totals	2906	10.352	2,245	5.388	1,815	4.490
14.	a	1	.009			1	.007
	b	1	.015			1	.007
	c	1	.012			1	.007
	d	1	.015			1	.008
	e	1	.011			1	.007
	Setup		.022				.023
	Totals	5	.084	5	.056	5	.059

[+] Due to the nature of the lexicographic serial approach, computation times for individual problems cannot be broken out.

[#] Feasibility generator time plus setup time.

Table J—*Continued*

		Algorithm F′		Algorithm G′⁺		Algorithm H′	
Problem	Objective Function Identifier	Number of Nodes	Total Time	Number of Nodes	Total Time	Number of Nodes	Total Time

B. Randomly generated 500 variable
problems with 10 agents and 50 tasks with $K = 5$

15.	a	51	.181			51	.112
	b	156	.583			71	.156
	c	162	.517			49	.108
	d	94	.260			23	.051
	e	60	.179			25	.550
	Setup		.040				.129
	Totals	523	1.760	512	1.251	219	1.106
16.	a	49	.192			49	.162
	b	25	.111			1	.006
	c	139	.617			69	.228
	d	238	1.202			143	.472
	e	60	.230			1	.006
	Setup		.050				.124
	Totals	511	2.402	313	1.160	263	.998
17.	a	211	.802			211	.780
	b	708	2.847			301	1.130
	c	388	1.836			355	1.311
	d	222	.915			363	1.342
	e	839	4.103			599	2.224
	Setup		.098				.124
	Totals	2,368	10.601	1,589	6.032	1,829	6.911
18.	a	89	.453			89	.312
	b	59	.236			41	.144
	c	52	.215			35	.123
	d	842	4.003			379	1.330
	e	110	.653			79	.277
	Setup		.102				.112
	Totals	1,152	5.662	1,781	6.228	623	2.298

Table J—*Continued*

Problem	Objective Function Identifier	Algorithm F′		Algorithm G′⁺		Algorithm H′	
		Number of Nodes	Total Time	Number of Nodes	Total Time	Number of Nodes	Total Time
C. Randomly generated 250 variable problems with 5 agents and 50 tasks with $K = 2$							
19.	a	19	.075			19	.069
	b	10	.051			19	.082
	Setup		.011				.007
	Totals	29	.137	29	.093	38	.158
20.	a	63	.219			63	.220
	b	104	.453			85	.340
	Setup		.011				.012
	Totals	167	.683	117	.317	148	.572
21.	a	73	.312			73	.315
	b	59	.299			87	.374
	Setup		.015				.015
	Totals	132	.626	118	.369	160	.704
22.	a	35	.168			35	.163
	b	114	.502			139	.629
	Setup		.012				.013
	Totals	149	.682	155	.591	174	.805
23.	a	39	.173			39	.168
	b	24	.109			43	.194
	Setup		.012				.011
	Totals	63	.294	63	.214	82	.373

parameterizations, while Algorithm G′ tends to be best for "minor" parameterizations.

D. The Generalized Assignment Problem with a Continuous Objective Function Parameterization

In this section we consider problem 6). Our goal is to find optimal solutions for all possible weightings of two different objective functions.

Due to the relatively poor results for Algorithm F′ in Section C, we did not code an approach using the serial method with an advanced initial separation. An algorithm corresponding to G′ was not developed because of

Table K. Ratios of Computation Times for Algorithms F', G', H' with H'
Scaled to 1

Problem	Number of Variables	Algorithm F'	Algorithm G'	Algorithm H'
11.	250	1.50	1.00	1.
12.	250	2.46	2.27	1.
13.	250	2.35	1.23	1.
14.	250	1.44	.95	1.
15.	500	1.60	1.13	1.
16.	500	2.41	1.16	1.
17.	500	1.53	.87	1.
18.	500	2.45	2.70	1.
19.	250	.87	.59	1.
20.	250	1.19	.55	1.
21.	250	.89	.52	1.
22.	250	.85	.73	1.
23.	250	.79	.57	1.

the inherent difficulty involved in choosing values of θ at which to solve a
relaxation at a given node. Only the approach corresponding to Algorithm H'
was coded. Algorithm H'' is similar to Algorithm D'' in Chapter V, Section D,
with the exception of using Algorithm E instead of Algorithm A in step 2 and
replacing $LB(\theta)$ by $UB(\theta)$. The reader is referred to Chapter V, Section D, for
an explanation of the reasoning behind this approach.

Test problems were generated per Ross and Soland [1975] in precisely
the same manner as given in Section A. The f_{ij} coefficients were generated
randomly from a $U[10,50]$ distribution that had a 50 percent correlation
coefficient with the corresponding c_{ij}.

Computational results are shown in Table L. Note that just as with the
0–1 knapsack results of Chapter V, Section D, the number of problems, (Q_θ),
solved is slightly less than twice the number of optimal solutions found over
$\forall \theta \varepsilon [0,1]$. This is further substantiation of the effectiveness of the method
for solving the continuum of problems (Q_θ) $\forall \theta \varepsilon [0,1]$.

Table L. Algorithm H″ for the Generalized Assignment Problem with a Continuous Objective Function (Time in seconds excluding I/O on an IBM 360/91)

Problem	Agents	Tasks	Variables	No. of Nodes	Total Time	No. of Values of θ Solved	Total No. of Optimal Solutions for (Q_θ) $\forall$ θ ε [0,1]
24.	5	50	250	682	2.14	27	14
25.	5	50	250	21	.32	21	11
26.	5	50	250	301	.92	27	14
27.	5	50	250	2,399	7.06	11	6
28.	5	50	250	1,721	5.72	14	9
29.	5	50	250	1,313	3.88	16	9
30.	5	50	250	29	.42	29	15

VII. The Parametric Capacitated Facility Location Problem

Consider the problem:

$$\min_{\substack{x_{ij} \geq 0 \\ y_i = 0,1}} \sum_{i \varepsilon I} \sum_{j \varepsilon J} c_{ij} x_{ij} + \sum_{i \varepsilon I} f_i y_i$$

(P)

$$\sum_{j \varepsilon J} x_{ij} \leq S_i y_i \qquad \forall \, i \, \varepsilon \, I$$

$$\sum_{i \varepsilon I} x_{ij} = D_j \qquad \forall \, j \, \varepsilon \, J$$

where $c_{ij} \geq 0, f_i \geq 0, S_i > 0$, and $D_j > 0$. This problem is often referred to as the capacitated facility location problem. The index set I denotes a collection of potential facilities (plants or warehouses), and J denotes a collection of customers that are to be serviced by the facilities. The maximum throughput for facility i is S_i, and D_j is the demand attributable to customer j. The cost, c_{ij}, is the transportation cost for supplying one unit of demand to customer j from facility i. The fixed cost, f_i, is the cost of opening facility i.

A great deal of research has been devoted to this particular problem class. An exhaustive survey on the subject is given in A. N. El-Shafei and K. B. Haley [1974]. Geoffrion [1975] presents an up-to-date catalog of current solution methodologies and computer codes available for solving various formulations of the facility location problem. Quite recently U. Akinc and B. M. Khumawala (AK) [1977] devised an efficient branch and bound algorithm that has greatly reduced computation times for (P) over existing algorithms.

In Section A, we shall present a branch and bound algorithm for (P) that is comparable in computation time to AK's algorithm for most test problems and improves upon their results for other test problems. In Section B, we shall present an approach for solving the general parametric capacitated facility location problem:

7) For $k = 1, \ldots, K$ solve:

$$\min_{\substack{x_{ij} \geq 0 \\ y_i = 0,1}} \sum_{i \varepsilon I} \sum_{j \varepsilon J} (c_{ij} + d_{ijk}) x_{ij} + \sum_{i \varepsilon I} (f_i + g_{ik}) y_i$$

70

(P_k)
$$\sum_{j \varepsilon J} x_{ij} \leqslant (S_i + T_{ik}) y_i \qquad \forall \; i \, \varepsilon \, I$$

$$\sum_{i \varepsilon I} x_{ij} = (D_j + E_{jk}) \qquad \forall \; j \, \varepsilon \, J$$

where d_{ijk}, g_{ik}, T_{ik}, and E_{jk} are scalars.

A. An Algorithm for the Capacitated Facility Location Problem

Consider problem (P). We shall begin the analysis of this problem by presenting basic, well-known results, and then some new results will be proved. After this, an efficient algorithm will be stated, and computational results will be cited.

G. Sà [1969] observed that in the continuous relaxation of (P), namely $(\bar{P})$, where $0 \leqslant y_i \leqslant 1$ replaces $y_i = 0,1$, it is possible to substitute the y_i variables out of the problem. That is, let $y_i = \sum_{j \varepsilon J} x_{ij} \, S_i$. This may be done since for any optimal solution $(\bar{x}, \bar{y})$ to $(\bar{P})$, we necessarily must have $\sum_{j \varepsilon J} \bar{x}_{ij} = S_i \bar{y}_i \; \forall \; i \, \varepsilon \, I$. The substitution is carried out by replacing the supply constraints by $\sum_{j \varepsilon J} x_{ij} \leqslant S_i \; \forall \; i \, \varepsilon \, I$, and the objective function by $\min \sum_{i \varepsilon I} \sum_{j \varepsilon J} (c_{ij} +$
$$x_{ij} \geqslant 0$$
$f_i / S_i) x_{ij}$. It is obvious that the resulting problem is a transportation problem. Hence, $(\bar{P})$ may be solved using a transportation or network algorithm rather than a linear programming algorithm, thus allowing for significant savings in computation time.

L. B. Ellwein [1970] gives the following result. Let $(P_A) \triangleq (P | y_i = 1 \; \forall \; i \, \varepsilon \, I)$.

Theorem 14. If $v(P_A | y_{i_o} = 0) - v(P_A) \geqslant 0$, then the optimal solution value for (P) is not affected by setting $y_{i_o} = 1$.

Basically, this result states that if the added transportation cost incurred by closing facility i_o is greater than or equal to the fixed cost of opening facility i_o, then facility i_o may be fixed open in an optimal solution to (P).

AK [1977] give the following result. Let $T_{i_o} \triangleq \{j | c_{i_o j} = \min_i c_{ij}\}$ and for $\forall \; j \, \varepsilon \, T_{i_o}$ let $\lambda_j \triangleq \min_{i \neq i_o} c_{ij}$. The index set T_{i_o} is that set of customers j for which facility i_o supplies the demand for the least cost. The value $\lambda_j, j \, \varepsilon \, T_{i_o}$ is the second smallest transportation cost for customer j.

Theorem 15. If $\max_{0 \leqslant x_{i_o j} \leqslant D_j} \sum_{j \varepsilon T_{i_o}} (\lambda_j - c_{i_o j}) x_{i_o j} \geqslant f_{i_o}$, then the optimal solution value for (P) is not affected by setting $y_{i_o} = 1$.

The knapsack problem in the theorem statement gives a measure of the transportation cost savings that can be realized if facility i_o is opened. If this saving is larger than the fixed cost f_{i_o}, then the facility may be pegged open.

AK show that Theorem 14 dominates Theorem 15. Specifically, if y_{i_o} can be pegged to 1 via Theorem 15, then it can always be pegged to 1 via Theorem 14, but not vice versa. Both theorems allow one to peg facilities open if a particular test is passed. For the Ellwein test, a transportation problem is solved for facility i. The AK test on the other hand only requires a continuous knapsack to be solved for each facility i. Computationally then, the (AK) test is more attractive even though it is a weaker test.

We now present a result that tightens the constraint structure of (P). Let $Q_{i_o} \triangleq \{j \,|\, c_{i_o j} = \min_i c_{ij} \text{ and } c_{i_o j} \neq \min_{i \neq i_o} c_{ij}\}$, and $L_{i_o} \triangleq \min \{ \sum_{j \in Q_{i_o}} D_j, S_{i_o}\}$. If $Q_{i_o} = \phi$, then define $\sum_{j \in Q_{i_o}} D_j$ to be zero. The index set Q_{i_o} denotes those customers j for which facility i_o has the *unique* smallest transportation cost over all facilities. Let (x^*, y^*) be an optimal solution for (P).

Theorem 16. If the constraint $\sum_{j \in J} x_{i_o j} \leq S_{i_o} y_{i_o}$ is replaced by $L_{i_o} y_{i_o} \leq \sum_{j \in J} x_{i_o j} \leq S_{i_o} y_{i_o}$, the optimal solution to (P) will not be affected.

Proof. If $y_{i_o}^* = 0$ the two constraints are equivalent. For $y_{i_o}^* = 1$ there are two cases: a) $\sum_{j \in J} x_{i_o j}^* = S_{i_o}$ and b) $\sum_{j \in J} x_{i_o j}^* < S_{i_o}$. If a) holds, the two constraints are equivalent. Only case b) remains. We will show that $\sum_{j \in J} x_{i_o j}^* \geq L_{i_o}$. We assume $\sum_{j \in J} x_{i_o j}^* < L_{i_o}$. Recall that $L_{i_o} = \min \{ \sum_{j \in Q_{i_o}} D_j, S_{i_o}\}$. For some $j_o \in Q_{i_o}$ we must have $x_{i_o j_o}^* < D_{j_o}$, since otherwise we would have $\sum_{j \in Q_{i_o}} x_{i_o j}^* = \sum_{j \in Q_{i_o}} D_j \geq L_{i_o}$.

So for $x_{i_o j_o}^* < D_{j_o}$ $\exists$ an i_1 such that $x_{i_1 j_o}^* > 0$. Now by transferring one unit of flow from $x_{i_1 j_o}^*$, we clearly remain feasible, but since $j_o \in Q_{i_o}$ we have $c_{i_o j_o} < c_{i_1 j_o}$, and so the modified solution is not only feasible but it also has a smaller objective function value. Contradiction. $\quad \|$

This result allows us to put a lower bound on the throughput of any facility in an optimal solution to (P). By adding this lower bound on throughput, we theoretically tighten the primary relaxation $(\bar{P})$. However, since $\sum_{j \in J} \bar{x}_{ij} = S_i \bar{y}_i \ \forall \ i \in I$, and since $L_i \leq S_i \ \forall \ i \in I$, we see that this added restriction on throughput has no effect on $(\bar{P})$. Nonetheless, it can be used to advantage for other relaxations that are tighter than $(\bar{P})$.

We shall now give a relaxation that is at least as strong as $(\bar{P})$ in objective value and that generally yields stronger penalties. Consider the following Lagrangean relaxation where the demand constraints are placed in the objective function:

$$\min_{\substack{D_j \geqslant x_{ij} \geqslant 0 \\ y_i = 0,1}} \sum_{i \in I} \sum_{j \in J} c_{ij} x_{ij} + \sum_{i \in I} f_i y_i + \sum_{j \in J} \bar{\lambda}_j \left(D_j - \sum_{i \in I} x_{ij} \right)$$

$$\sum_{j \in J} x_{ij} \leqslant S_i y_i \qquad \forall \ i \varepsilon \ I,$$

or equivalently

$$\sum_{j \in J} \bar{\lambda}_j D_j - \max_{\substack{D_j \geqslant x_{ij} \geqslant 0 \\ y_i = 0,1}} \sum_{i \in I} \sum_{j \in J} (\bar{\lambda}_j - c_{ij}) x_{ij} - \sum_{i \in I} f_i y_i$$

$(LGR_1^{\bar{\lambda}})$
$$\sum_{j \in J} x_{ij} \leqslant S_i y_i \qquad \forall \ i \varepsilon \ I.$$

This is a valid relaxation and it is easily seen that $v(P) \geqslant v(LGR_1^{\bar{\lambda}}) \geqslant v(\bar{P})$ if optimal dual multipliers, $\bar{\lambda}$, from $(\bar{P})$ are used. We mention that if a transportation algorithm is used to solve $(\bar{P})$ (the y_i's having been substituted out of the problem), then the dual multipliers for the LP formulation of $(\bar{P})$ must be recovered from the optimal dual multipliers of the transportation formulation. The recovery is not difficult to accomplish, consisting only of elementary transformations for a number of special cases. These transformations yield $\bar{\lambda}$.

Unfortunately, in practice it is sometimes the case that $v(LGR_1^{\bar{\lambda}}) = v(\bar{P})$. Thus, the Lagrangean relaxation may be no stronger than the continuous relaxation. However, there are at least three different calculations that generally tighten the relaxation measurably.

The first is to find a good set of λ's. Specifically, set $\tilde{\lambda}_j = \max \{\bar{\lambda}_j, \tilde{\lambda}_j\}$, $1 \leqslant j \leqslant m$ where $\tilde{\lambda}_j$ is the second smallest transportation cost (including ties) for customer j over all available facilities. It is easy to see that this "improved" set of dual multipliers tightens the relaxation. For example, suppose $\bar{\lambda}_j < \tilde{\lambda}_j$. Then the constant term in (LGR_1^{λ}) increases in value from $\bar{\lambda}_j D_j$ to $\tilde{\lambda}_j D_j$ while the term $(\bar{\lambda} - c_{ij})$ is greater than 0 for at most one facility i. Since the corresponding variable $x_{ij} \leqslant D_j$, we see that $v(LGR_1^{\lambda})$ will increase by $(\tilde{\lambda}_j D_j - \bar{\lambda}_j D_j)$ and decrease by no more than $(\tilde{\lambda}_j - c_{ij}) x_{ij} - (\bar{\lambda}_j - c_{ij}) x_{ij}$. Hence, $v(LGR_1^{\lambda})$ is at least as tight with the improved multipliers.

The second improvement is the addition of the constraints: $L_i y_i \leqslant \sum_{j \in J} x_{ij} \ \forall \ i \varepsilon \ I$ to $(LGR_1^{\bar{\lambda}})$ where L_i is the lower bound on throughput for facility i as given in Theorem 16.

The third improvement is to append the constraint: $\sum_{i \in I} S_i y_i \geqslant \sum_{j \in J} D_j$ to (LGR_1^{λ}). This constraint forces feasibility of any facility design by requiring that sufficient facilities are opened to handle the total demand. Of course, the customer demand constraints may still be violated in (LGR_1^λ) since they have been relaxed.

By combining all three improvements, we have the following tighter relaxation:

$$\sum_{j\in J} \hat{\lambda}_j D_j - \max_{\substack{0 \leq x_{ij} \leq D_j \\ y_i = 0,1}} \sum_{i\in I}\sum_{j\in J}(\hat{\lambda}_j - c_{ij})x_{ij} - \sum_{i\in I} f_i y_i$$

(LGR_2^λ)

$$L_i y_i \leq \sum_{j\in J} x_{ij} \leq S_i y_i \qquad \forall\, i \,\varepsilon\, I$$

$$\sum_{i\in I} S_i y_i \geq \sum_{j\in J} D_j.$$

While this relaxation may appear to be difficult to solve efficiently, we shall show that it is not. By projecting on the space of y variables, we see that the optimization over the x variables can be carried out independently. Clearly if $\hat{y}_{i_o} = 0$, then $\hat{x}_{i_o j} = 0 \ \forall\, j\,\varepsilon\, J$. If $\hat{y}_{i_o} = 1$, then $\hat{x}_{i_o j} = \bar{x}_{i_o j} \ \forall\, j\,\varepsilon\, J$ where $\bar{x}_{i_o j}$ is an optimal solution to:

$$v(i_o, \hat{\lambda}) = \max_{0 \leq x_{i_o j} \leq D_j} - f_{i_o} + \sum_{j\in J}(\hat{\lambda}_j - c_{i_o j})x_{i_o j}$$

$$L_{i_o} \leq \sum_{j\in J} x_{i_o j} \leq S_{i_o}.$$

Hence, for each $i \,\varepsilon\, I$ a continuous knapsack is solved to obtain $v(i, \hat{\lambda})$. Then the following problem is solved over y:

$$\max_{y_i = 0,1} \sum_{i\in I} v(i, \hat{\lambda}) y_i$$

(F)

$$\sum_{i\in I} S_i y_i \geq \sum_{j\in J} D_j.$$

Problem (F) is a 0–1 knapsack problem in y that may be solved efficiently by the methods of Chapter V. Let $\tilde{y}$ denote an optimal solution of (F). We note that generally the knapsack has fewer than $|I|$ free variables, since if $v(i, \hat{\lambda}) \geq 0$ we may set $\tilde{y}_i = 1$ in any optimal solution to (F). Thus, to solve (LGR_2^λ) we first solve a continuous knapsack in $x_{i_o(\cdot)}$ for each facility i_o. Then, we use the solution values for these knapsacks to solve a 0–1 knapsack in y.

Note that penalties may be easily calculated for the y variables. For example, suppose $\tilde{y}_i = 1$ in (F). A penalty for $y_i = 0$ may be found by solving $(F|y_i = 0)$ or $(\bar{F}|y_i = 0)$ where $(\bar{F})$ is the continuous relaxation of (F). Thus $(\bar{F})$ is a continuous knapsack. Clearly, at a given node in a branch and bound tree if $v(\bar{F}|y_i = 0) \geqslant z^*$, where z^* is the current incumbent value, we may set $y_i = 1$ in all successors of that node. Further, these continuous knapsack solutions may be used in choosing the next branch variable at a node. Specifically, one may order the y_i variables in decreasing order of $P_i \triangleq$ max $\{0, |v(\bar{F}|y_i = 1 - \tilde{y}_i) - v(F|y_i = \tilde{y}_i)|\}$. The branch variable is chosen to be $i_o = \{i | \max_i P_i\}$. If a LIFO priority scheme is used, then one would place $(CP|y_{i_o} = 1 - \tilde{y}_{i_o})$ and $(CP|y_{i_o} = \tilde{y}_{i_o})$ in the candidate list in that order. Thus, the $y_{i_o} = \tilde{y}_{i_o}$ branch would be examined first.

We shall now state an efficient algorithm for (P). Let $K_0(K_1)$ be the index set for y_i set to $0(1)$. Let $K_2 \triangleq I - K_0 - K_1$.

Algorithm I:

1. Set $K_0 = K_1 = \phi$, $K_2 = I$, and let z^* be a large number.

2. For $\forall\ i\ \varepsilon\ K_2$ perform the pegging test of Theorem 15. If successful, set $y_i = 1$ and set $K_1 = K_1 \cup \{i\}$.

3. For $\forall\ i\ \varepsilon\ K_2$ perform the pegging test of Theorem 14. If successful, set $y_i = 1$ and set $K_1 = K_1 \cup \{i\}$.

4. Initialize the candidate list to consist of $(P|y_i = 1\ \forall\ i\ \varepsilon\ K_1)$. Call this problem (CP) and go to 9.

5. Stop if the candidate list is empty: if there exists an incumbent, then it must be optimal in (P), otherwise (P) has no feasible solution.

6. Select a problem (CP) from the candidate list using a LIFO rule. Reset K_0, K_1 to coincide with the restrictions of (CP). If $BND(CP)$ exists and if $BND(CP) \geqslant z^*$, go to 5. (See step 18 for definition of $BND(\cdot)$.)

7. For $\forall\ i\ \varepsilon\ K_2$ perform the pegging test of Theorem 15. If successful, set $y_i = 1$ and set $K_1 = K_1 \cup \{i\}$.

8. For $\forall\ i\ \varepsilon\ K_2$ if $\sum\limits_{l\varepsilon K_1 \cup K_2 - \{i\}} S_l < \sum\limits_{j\varepsilon J} D_j$, then set $y_i = 1$ and set $K_1 = K_1 \cup \{i\}$.

9. Let (CP) be replaced by $(CP|y_i = 0, i\ \varepsilon\ K_0; y_i = 1, i\ \varepsilon\ K_1)$. Solve $(\overline{CP})$ as a transportation problem, getting a solution $(\bar{x}, \bar{y})$.

10. If $(\overline{CP})$ is infeasible, go to 5.

11. If $v(\overline{CP}) \geqslant z^*$, go to 5.

12. If $(\bar{x}, \bar{y})$ is feasible for (CP), go to 21.

13. If $v(\overline{CP}) + \sum\limits_{i:0 < \bar{y}_i < 1} (1 - \bar{y}_i)f_i < z^*$, go to 22.

14. Solve (LGR_2^{λ}) corresponding to (CP), getting a solution (x^*, y^*).

15. If $v(LGR_2^{\lambda}) \geqslant z^*$. go to 5.

16. Calculate penalties P_i by solving $(\overline{F}|y_i = 1 - y_i^*) \; \forall \; i \; \varepsilon \; K_2$.

17. For $\forall \; i \; \varepsilon \; K_2$ if $P_i + v(LGR_2^{\lambda}) \geqslant z^*$, set $y_i = y_i^*$ and set $K_{y_i^*} = K_{y_i} \cup \{i\}$. If a y_i has been pegged to 0 in this step, go to 19. Otherwise if a y_i has been pegged to 1 add $(CP|y_i = 0, i \; \varepsilon \; K_0 ; y_i = 1, i \; \varepsilon \; K_1)$ to the candidate list and go to 6. If no variables have pegged to 0 or 1, go to 18.

18. Find $i_o = \{i | \max\limits_{i \varepsilon K_2} P_i\}$. If there is a tie, break it arbitrarily. Add the problems $(CP|y_{i_o} = 1 - y_{i_o}^*)$ and $(CP|y_{i_o} = y_{i_o}^*)$ to the candidate list. Associate the value $BND(CP|y_{i_o} = 1 - y_{i_o}^*) = P_{i_o} + v(LGR_2^{\lambda})$ with $(CP|y_{i_o} = 1 - y_{i_o}^*)$ and go to 6.

19. For $\forall \; i \; \varepsilon \; K_2$ if $\sum\limits_{l \varepsilon K_1 \cup K_2 - \{i\}} S_l < \sum\limits_{j \varepsilon J} D_j$, set $y_i = 1$ and set $K_1 = K_1 \cup \{i\}$.

20. Replace the current (CP) by $(CP|y_i = 0, i \; \varepsilon \; K_0 ; y_i = 1, i \; \varepsilon \; K_1)$. For $\forall \; j \; \varepsilon \; J$ let $\hat{\lambda}_j = \max \{\hat{\lambda}_j, c_{i_2 j}\}$. Go to 14.

21. An improved feasible solution has been found. Set $z^* = v(\overline{CP})$ and record the associated solution $(\bar{x}, \bar{y})$ as the new incumbent. Go to 5.

22. An improved feasible solution has been found. Set $z^* = v(\overline{CP}) + \sum\limits_{i:0 < \bar{y}_i < 1} (1 - \bar{y}_i)f_i$. If $0 < \bar{y}_i < 1$, set $\bar{y}_i = 1$, and record $(\bar{x}, \bar{y})$ as the new incumbent. Go to 14.

In step 2, we perform the continuous knapsack pegging test of Theorem 15 before the pegging test of Theorem 14. While the former test is weaker than the latter, computation time is generally much smaller. In step 7, the pegging test of Theorem 15 is used before the primary relaxations are solved. This is done since the pegging test is independent of the primary relaxation solution, and in fact does not even depend on an incumbent value, z^*. Hence, if a variable can be pegged to 1 via this test, the primary relaxations will be tightened. In step 8, a simple conditional feasibility test is invoked. This test assumes that y_{i_o} is set to 0. If total demand cannot be satisfied by opening all facilities in $K_2 - \{i_o\}$, then y_{i_o} may be pegged to 1. In step 9, all $y_i, i \; \varepsilon \; K_2$ are substituted out of $(\overline{CP})$, and the objective function and constraints are modified accordingly (as explained earlier). A transportation or network algorithm is then used to solve $(\overline{CP})$. In step 13, the fractional $\bar{y}_i$'s in the transportation solution are rounded up (thus assuring integer feasibility) in an attempt to generate an improved incumbent. Of course, even if this rounded solution results in an improved incumbent, the current candidate problem is not fathomed. In step 17, a simple conditional test is used to try to peg facilities open or closed. If a facility is pegged closed, then the conditional feasibility test in step 18 is invoked, since by pegging a variable to 0 this test is strengthened. Then dual multipliers are improved in step 19, and the Lagrangean relaxation of step 14 is solved. Note that the transportation relaxation is bypassed in this case. This is done since, generally, the transportation relaxation takes much more computation time than does the Lagrangean relaxation. If no variables are pegged to 0 or 1 in step 17, a branch must be made. The branching criteria is to choose that variable with the largest penalty, P_i. The two resulting problems are placed in the candidate list such that the most promising branch is examined first under the LIFO rule of step 6. In step 18, note that a bound of $P_i + v(LGR_2^\lambda)$ is associated with $(CP|y_{i_o} = 1 - y_{i_o}^*)$. When this candidate problem is selected from the candidate list, this bound is compared with z^* to see if it may be "automatically" fathomed.

The test problems used were a subset of the test problems of A. A. Kuehn and M. J. Hamburger [1963] and Ellwein [1970]. These problems have served as bench marks for researchers studying the capacitated facility location problem. As previously mentioned, AK have devised a new branch and bound algorithm that has dramatically reduced computation times for this class of test problems.

We mention that in our computer implementation we have not incorporated Theorem 16 and $(\bar{F})$ was used instead of (F) in the calculation of $v(LGR_2^\lambda)$. In Table M, our results are compared with the published results of AK. Care should be taken in comparing times for at least two reasons. First, different computers were used. Second, and more importantly, different transportation codes were used. Since transportation code time often accounted for over 90 percent of the total computation time in our implementation, a more realistic comparison would be the number of transportation problems solved. In support of this, we note that AK used an out-of-kilter code for certain problems and a primal-dual code for others. Both of these codes had a complete reoptimization capability. Our code, on the other hand, had only a limited reoptimization capability. Specifically, given an optimal solution to $(\overline{CP})$, reoptimization was only possible when a variable y_i was set to 1. Reoptimization was not possible when y_i was set to 0. As further evidence of the reoptimization capability, AK solved a 25 facility by a 50 customer problem using the test of Theorem 14 and making one branch in about .2 seconds. With our transportation code these same computations took over 5 seconds. Generally, we observed that for easy problems (15 or less transportation problems solved) the results were quite similar. However, for more difficult problems, our algorithm generally required the solution to fewer transportation problems.

Table M. Comparison of AK Algorithm and Algorithm I
(Time in seconds excluding I/O)

Problem	Supplies	Fixed Costs	No. of Facilities	No. of Customers	AK Algorithm (IBM 370/165)		Algorithm I (IBM 370/168)	
					No. of Transp. Problems Solved	Total Time	No. of Transp. Problems Solved	Total Time
1.	5,000	7,500	16	50	14	10.2	14	7.6
2.	5,000	12,500	16	50	7	9.2	1	1.5
3.	5,000	17,500	16	50	7	9.3	9	6.8
4.	5,000	25,000	16	50	7	9.6	13	10.8
5.	10,000	17,500	16	50	65	19.7	9	6.6
6.	15,000	7,500	16	50	4	.2	3	1.5
7.	15,000	12,500	16	50	9	.4	2	1.7
8.	15,000	17,500	16	50	183	38.6	8	5.3
9.	15,000	25,000	16	50	133	34.4	2	2.9
10.	5,000	7,500	25	50	4	.2	16	18.4
11.	5,000	12,500	25	50	—	>120	15	18.1
12.	15,000	7,500	25	50	10	.8	3	2.2

Table M—*Continued*

Problem	Supplies	Fixed Costs	No. of Facilities	No. of Customers	AK Algorithm (IBM 370/165)		Algorithm I (IBM 370/168)	
					No. of Transp. Problems Solved	Total Time	No. of Transp. Problems Solved	Total Time
13.	15,000	12,500	25	50	82	7.8	17	9.2
14.	1,000–5,000	15,000–40,000	15	45	43	12.6	5	1.8
15.	1,500–7,500	22,500–60,000	15	45	20	4.2	15	8.1
16.	5,000	12,500	10	20	NA	NA	19	1.4
17.	5,000	13,750	10	20	NA	NA	21	1.1
18.	5,000	15,000	10	20	NA	NA	23	1.3
19.	5,000	25,000	10	20	NA	NA	44	2.1

B. An Interactive Approach for the Parametric Capacitated Facility Location Problem

In this section, we consider the general parametric capacitated facility location problem, 7). The approach that we shall propose applies generally to problem 7), however, the test problems that we have studied vary only one of the possible parameters at a time. Specifically, we have tested problems with varying demand only, varying fixed costs only, and varying transportation costs (c_{ij}) only. Because of the computer cost involved in solving capacitated facility location problems, we chose not to run test problems for the continuous parameterizations in the objective function and in the right-hand side.

The plan of this section is to investigate the problem dependent techniques of Chapter III, and the factors affecting the scheduling of solution priorities in Chapter IV. Then an interactive approach for solving problem 7) is proposed, and computational experience is cited.

Reduction techniques for problem 7) can be quite effective. The pegging tests of theorems 14 and 15 should be used for each (P_k). Note that if a parameterization involves only the fixed costs, then both pegging test calculations need only be done once. This follows since the calculations remain the same, regardless of the value of the fixed costs. All that must be done is to compare the calculated figure with the fixed cost for each $k = 1, \ldots, K$. For other parameterizations the test of Theorem 15 may be done separately for each $k = 1, \ldots, K$, since all that is involved is the solution of a continuous knapsack for each facility i for each $k = 1, \ldots, K$. The test of Theorem 14, on the other hand, involves solving a transportation problem for each facility i for each $k = 1, \ldots, K$. To take advantage of efficient reoptimization techniques for this test, one could solve the K transportation problems for y_i, say, as a group, reoptimizing the optimal tableau from $k = 1$ for $k = 2$, and so on. This reoptimization may reduce total computation time for this test.

Feasibility recovery techniques have turned out to be important for this class of problems. In fact, for about one half of the parameterizations tested, the feasibility recovery technique found an optimal solution for the next problem in the parameterization. With the use of this solution value as an upper bound, it was possible to terminate the branch and bound search in some problems without any branching whatsoever. We shall explain this behavior in detail later. Clearly, if the parameterization involves only the objective function, there are at least two methods for generating a feasible solution for (P_2) from an optimal solution, $(x^*,y^*)_1$ to (P_1). First, since $(x^*,y^*)_1$ is feasible in (P_2), we may simply cost out that solution using the objective function for (P_2). Second, by using an optimal design of facilities for (P_1), we may solve the associated transportation problem with the objective function for (P_2). This latter method was used in our computer implementation. If the parameterization involves the right-hand side and $(x^*, y^*)_1$ remains feasible for (P_2), one may also solve the associated transportation problem using the optimal design of facilities for (P_1). If $(x^*, y^*)_1$ is not feasible in (P_2), then one may, for example, selectively open enough facilities in addition to the open facilities in the optimal solution to (P_1) until feasibility is assured, that is, $\sum_{i: y_i = 1} S_i \geqslant \sum_{j \in J} D_j$. Then the associated transportation problem can be solved.

Bounding problem reoptimization techniques can also be important. In general, at a fathomed node for (P_1), one could take the optimal transportation tableau and use it as a starting tableau for optimizing the corresponding candidate problem for (P_2). If an interactive procedure is used, one could store the optimal tableau for the candidate problem on disk or tape and then read this information back into core when the candidate problem for (P_2) at that node is considered.

Finally, wide range bounding techniques may be employed at a given node. There are at least two possibilities. First, (LGR_2^λ) may be solved for (P_2) using the $\hat{\lambda}$ from the transportation problem solved at the node for (P_1). Second, all constraints could be absorbed into the objective function using the appropriate dual multipliers from the transportation problem solved for (P_1) at the node. However, since both of these relaxations are of moderate computational expense, the former approach (which is stronger) should probably be used.

We now turn to the factors affecting the scheduling of solution priorities. Tightness of the primary relaxation for a given (P_k) has been found to be dependent on one main characteristic. We assume for this analysis that the ratios f_i / S_i are equal for $\forall\ i\ \varepsilon\ I$. Let $(\bar{x}, \bar{y})$ be an optimal solution for $(\bar{P}_k)$. If $\sum_{i \in I} f_i \bar{y}_i$ is "close" to $\sum_{i \in I} f_i y_i^*$, where (x^*, y^*) is an optimal solution to (P_k), then the relaxation will be relatively tight. Conversely, if $\sum_{i \in I} f_i \bar{y}_i$ is much smaller than $\sum_{i \in I} f_i y_i^*$, then the relaxation will be relatively loose. Now if fixed costs are relatively high with respect to transportation costs, then facilities that are open in y^* will probably be operated close to their upper capacities. The corresponding relaxation will probably be tight. If, on the other hand, fixed costs are relatively low with respect to transportation costs, then facilities that are open in y^* will probably be operated at something less than their upper capacity. In this case, the continuous relaxation will probably be looser. Still another way to look at relaxation tightness is the difference between $\sum_{i \in I} S_i y_i^*$ and $\sum_{i \in I} S_i \bar{y}_i = \sum_{j \in J} D_j$. This difference is manifested cost wise in the difference between $\sum_{i \in I} f_i y_i^*$ and $\sum_{i \in I} f_i \bar{y}_i$. By using these intuitive relationships, one can often "predict" the tightness of the continuous relaxation as a function of the parameterization. We have the following relationships:

Type of parameterization	Tendency of continuous relaxation
Increase c_{ij}'s	Tighten
Increase f_i's	Loosen
Increase S_i's	Loosen if $\sum_{i \in I} S_i y_i^* - \sum_{j \in J} D_j$ is small; tighten if large
Increase D_j's	Loosen if $\sum_{i \in I} S_i y_i^* - \sum_{j \in J} D_j$ is small; tighten if large

Note that these relationships are also dependent on the relative importance of transportation costs that, in turn, are affected by the number of facilities open in an optimal solution. Thus total transportation cost is higher if fixed costs are high, since this implies that fewer facilities are open in an optimal solution. So we see that the transportation costs have an effect opposite that of fixed costs on the tightness of the relaxation. This is reflected in the display above.

The behavior of individual facilities in an optimal solution as a function of certain parameterizations is closely related to the analysis above. For example, suppose all fixed costs rise by the same amount. This has the qualitative effect of making the opening of each facility less attractive. Coupled with the fact that the total transportation cost for any given facility design remains the same for both sets of fixed costs, it is easy to see that this rise in fixed costs is equivalent to tightening an implicit constraint on the number of open facilities in an optimal solution. Similarly, if demands rise, then one would expect additional facilities to be opened. A rise in transportation costs would be accompanied by a tendency toward opening more facilities also. Of course for certain "local" parameterizations, such as an increase in one fixed cost or an increase in certain transportation costs, tendencies of specific variables can be identified more precisely.

The persistence of scratch trees for the parameterizations that we tested was generally very good. Such behavior would seem to be plausible when one considers the size of the decision space for our test problems. The number of facilities considered ranged from 10 to 25, while the number of continuous variables ranged from 200 to 1,250. With a decision space of dimension 25 or less, and the empirical observation that the majority of facilities are "important," it is plausible that they remain "important" for most parameterizations of interest. If this is the case, scratch trees should be rather stable since "important" variables are branched on first in our algorithm. This has been borne out empirically, and indeed this bodes well for using a serial approach with an initial separation gleaned from the previous (P_k).

We shall now present a method for solving problem 7). Due to the relatively high computer cost for solution of reasonable sized problems, we chose an interactive approach. That is, (P_1) was solved to optimality in one computer run, and then using information from this computer run, another run was made to solve (P_2). By using this interactive approach, closer control over computation time was possible, and a freer hand in experimentation was permitted.

Algorithm J:

1. Set $k = 1$. Solve (P_k) by Algorithm I, getting an optimal solution, $(x^*, y^*)_k$.

2. If $k > K$, stop. If (P_k) and (P_{k+1}) are relatively monotone, if their objective functions are identical, and if $(x^*, y^*)_k \, \varepsilon \, F(P_{k+1})$, then $(x^*, y^*)_k$ is optimal in (P_{k+1}), so let $k = k + 1$, and return to the beginning of this step.

3. Set $k = k + 1$. If $k > K$, stop.

4. Use a feasibility recovery technique on $(x^*, y^*)_{k-1}$ to find a good feasible solution for (P_k). Call it $(x^*, y^*)_k$ and its corresponding value z_k^*.

5. Invoke Algorithm I for (P_k) with the following modifications:
a) Replace step 19 by: "store the current index sets K_0, K_1, K_2, the incumbent $(x^*, y^*)_k$, and z_k^*, and return to step 6 of Algorithm J."
b) Replace step 5 by: "stop if the candidate list is empty: if there exists an incumbent, it is optimal in (P_k), return to step 2 of Algorithm J."

6. Use some modification of the branch and bound tree for (P_{k-1}) to form an initial separation for (P_k). Put this initial separation in the form of a candidate list and go to step 6 of Algorithm I. When the stop condition in step 5 of Algorithm I is satisfied, go to step 2 of Algorithm J.

In step 2, since problem 7) is a general parametric problem, we require identical objective functions. However, this can be relaxed for certain objective function changes. In step 4, the particular feasiblity recovery technique will depend on the type of parameterization. In step 5, a modification of Algorithm I is used to perform the pegging tests of theorems 14 and 15, and other pegging tests based on the Lagrangean relaxation (LGR_2^λ) for (P_k). This procedure continues until either fathoming occurs (in which case (P_k) has been solved), or until a point is reached where no more pegs can be made. When no more pegs can be made, control passes to step 6 of Algorithm J. In step 6, an initial separation based on the branch and bound tree of (P_{k-1}) is generated. The reasoning behind this approach is as follows.

Generally, the pegging tests of theorems 14 and 15 are quite effective in pegging facilities open at the root node. Hence, these tests are performed for each (P_k) in order to reduce the number of free facilities in the problem and, hopefully, to reduce the number of branches made during the branch and bound process. A modification of the branch and bound tree for (P_{k-1}) is used to form an initial separation because of the pegging procedure in step 5. The modification can best be explained by referring to a typical branch and bound tree for (P_1) (see figure below).

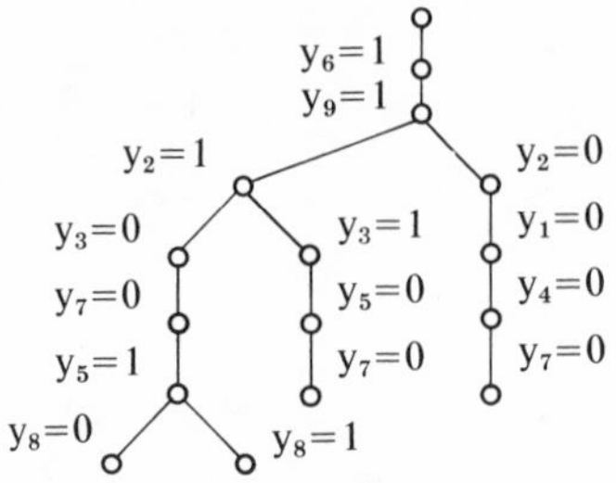

Branch and bound tree for (P_1)

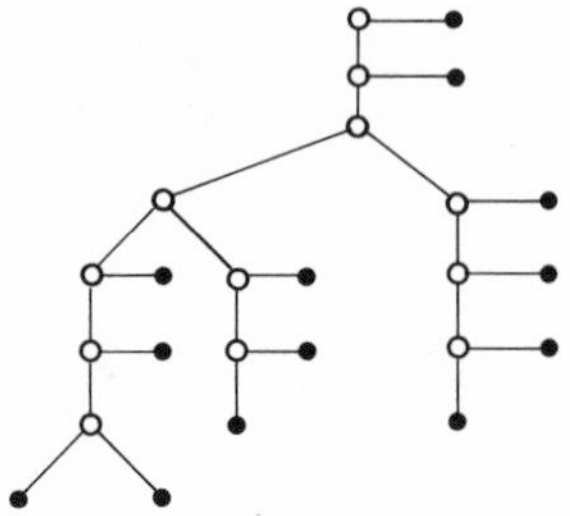

Frontier of fathomed nodes for (P_1)
(denoted by •)

Note that vertical lines refer to pegged variables. That is, the opposite branch is automatically fathomed. It is important to realize, however, that these opposite branches are part of the frontier of fathomed nodes as depicted in the figure above. If this full frontier were used as an initial separation for (P_2), the number of problems in the initial candidate list would be 13. However, by "collapsing out" the pegged variables from the tree, we may reduce the frontier to 4. See the figure below.

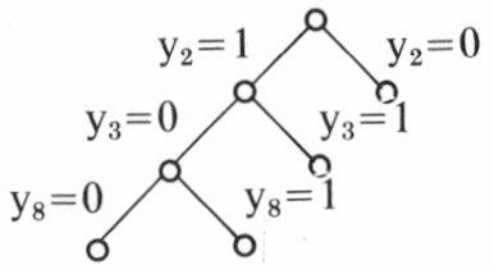

"Collapsed" branch and
bound tree for (P_1)

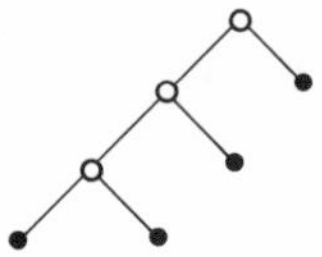

Frontier of fathomed
nodes for "collapsed"
tree (denoted by •)

While such a manipulation may appear contrived, we offer the following explanation. In the serial approach, which uses the frontier of fathomed nodes, there is an underlying supposition. It is that in a branch and bound tree if a node cannot be fathomed for say, (P_1), then it probably cannot be fathomed for a closely related problem, say, (P_2). Conversely, if a node is fathomed for (P_1), then it likely will be fathomed for (P_2) also. A variable that is pegged to a certain value can be thought of in an analogous manner. If y_i can be pegged to 1 in (P_1) at a given node, then it can probably be pegged to 1 in (P_2) also at that given node. Hence, instead of considering the full frontier of fathomed nodes for (P_1), we may eliminate the pegs from the tree and attempt to peg variables at the root node of (P_2) as in step 5 of Algorithm J.

We note that other methods for reducing the frontier of fathomed nodes, while still maintaining a "good" initial separation, are possible. However, we shall defer discussion of them to Chapter VIII.

As a final comment on Algorithm J, we submit that the ordering of the initial candidate list in step 6 can be very important. We present three orderings that have proven to be effective:

a) Order the candidate problems by the inverse order in which they were fathomed for the previous problem. That is, the candidate problem fathomed first is placed at the end of the list, so that under a LIFO rule, it will be examined first.

b) Order the candidate problems in decreasing order of their relaxation values for the previous problem. Thus, under a LIFO rule the most promising candidate problem will be examined first.

c) Modify the ordering in a or b by placing last in the list the candidate problem at which an optimal solution for the previous problem was found.

In our computational studies, ordering b coupled with modification c was generally the best of these orderings.

Test problems were taken from the problems in Table M. For example, problems 1, 2, 3, and 4 differ only in fixed costs, as do 6, 7, 8, and 9, and 16, 17, 18, and 19. Other problems were generated by increasing all customer demands by some percentage (that is, $D_j + 5$ percent), while still others were generated by increasing all transportation costs by some percentage ($c_{ij} + 10$ percent). Algorithm J was compared with the following traditional approach.

Algorithm K:

1. Set $k = 1$.

2. Solve (P_k) by Algorithm I, getting an optimal solution $(x^*, y^*)_k$.

3. Set $k = k + 1$. If $k > K$, stop.

4. Use a feasibility recovery technique on $(x^*, y^*)_{k-1}$ to find a good feasible solution for (P_k). Call it $(x^*, y^*)_k$ and its corresponding objective value z_k^*. Go to 2.

A comparison of the two algorithms is given in Table N. Note that the computation time includes the time for the Theorem 14 pegging tests, but that the number of transportation problems solved does not include the $(|I| + 1)$ transportation problems solved for this pegging test. We mention that the computation time for this test is much higher than it would be if a reoptimization capability were available in our transportation code. Because of this shortcoming, the savings realized by Algorithm J over Algorithm K are reduced. Note that in four of the eight parameterizations tested, only one transportation problem was required to solve each problem after (P_1). This was due to the fact that the solution generated by the feasibility recovery technique was an optimal solution, and that the Lagrangean relaxation penalties were so strong that many of the remaining free facilities were pegged

Table N. Comparison of Algorithms J and K for the Parametric Capacitated Facility Location Problem
(Time in seconds excluding I/O on an IBM 360/91. Both Algorithms used a suboptimality tolerance of .1 percent.)

Problem	Parameterization	No. of Facilities	No. of Customers	Size of Initial Candidate List	Algorithm J		Algorithm K	
					No. Transp. Problems Solved	Total Time	No. Transp. Problems Solved	Total Time
1.	Fix. Cst. 7,500	16	50	1	7	4.8	7	4.8
2.	Fix. Cst. 12,500	16	50	5	7	1.5	11	2.5
3.	Fix. Cst. 17,500	16	50	5	5	1.2	9	2.2
4.	Fix. Cst. 25,000	16	50	5	5	1.2	8	1.9
Totals					24	8.7	35	11.4
6.	Fix. Cst. 7,500	16	50	1	6	5.0	6	5.0
7.	Fix. Cst. 12,500	16	50	5	8	2.1	11	2.6
8.	Fix. Cst. 17,500	16	50	5	13	3.5	17	4.2
Totals					27	10.6	34	11.8
16.	Fix. Cst. 12,500	10	20	1	19	1.4	19	1.4
17.	Fix. Cst. 13,750	10	20	11	11	.7	21	1.1
18.	Fix. Cst. 15,000	10	20	11	13	.8	23	1.3
19.	Fix. Cst. 25,000	10	20	11	29	1.6	44	2.1
Totals					72	4.5	107	5.9
15.	Fix. Cst. f_i	15	45	1	11	5.9	11	5.9
15.	Fix. Cst. f_i + 10%	15	45	—	1	.3	1	.3
15.	Fix. Cst. f_i + 20%	15	45	—	1	.3	1	.2
15.	Fix. Cst. f_i + 30%	15	45	—	1	.2	1	.3
Totals					14	6.7	14	6.7

Problem	Parameterization	No. of Facilities	No. of Customers	Size of Initial Candidate List	Algorithm J		Algorithm K	
					No. Transp. Problems Solved	Total Time	No. Transp. Problems Solved	Total Time
6.	Demands D_j	16	50	1	6	5.0	6	5.0
6.	Demands $D_j + 5\%$	16	50	—	1	3.7	1	3.6
6.	Demands $D_j + 10\%$	16	50	—	1	3.7	1	3.7
Totals					8	12.4	8	12.3
15.	Demands D_j	15	45	1	11	5.9	11	5.9
15.	Demands $D_j + 5\%$	15	45	11	14	6.7	22	8.8
15.	Demands $D_j + 10\%$	15	45	11	11	5.7	11	5.7
Totals					36	18.3	44	20.4
1.	Transp. c_{ij}	16	50	1	7	4.8	7	4.8
1.	Transp. $c_{ij} + 10\%$	16	50	—	1	3.6	1	3.6
1.	Transp. $c_{ij} + 20\%$	16	50	—	1	3.5	1	3.6
Totals					9	11.9	9	12.0
6.	Transp. c_{ij}	16	50	1	6	5.0	6	5.0
6.	Transp. $c_{ij} + 10\%$	16	50	—	1	3.7	1	3.7
6.	Transp. $c_{ij} + 20\%$	16	50	—	1	3.7	1	3.7
Totals					8	12.4	8	12.4

open or closed. Then after these pegs were made, the corresponding La-grangean relaxation value bounded out.

In conclusion, we see that Algorithm J dominates Algorithm K. For various reasons mentioned throughout this chapter, it is reasonable to assume that this domination can only become more pronounced as more difficult problems are solved.

VIII. Extensions to the General PILP and Areas for Future Research

A. A Solution Method for the General PILP

In the preceding three chapters, we have analyzed PILPs for special problem classes. In this section, we extend what we have learned to the general PILP:

8) For $k = 1, \overset{i}{\ldots}, K$ solve:

$$\min (c + f_k)x$$

$$(P_k) \qquad (A + D_k)x \geqslant b + r_k$$

$$x_j \text{ integer}, j \, \varepsilon \, J$$

where f_k, D_k, and r_k are conformable with c, A, and b, respectively.

Our goal is to give prescriptions for solving the PILP in an efficient manner. In Chapter IV, three solution priorities were outlined: serial, lexicographically serial, and parallel. The reader will recall that only the first two priorities were considered in our computational studies, since the parallel approach is dominated by both of the other approaches. In this chapter, we shall reduce consideration to only one approach, namely, the serial.

The lexicographically serial approach (l.s.a.) has one inherent weakness. It is that individual (P_k)s are not solved to optimality in any preordained order. For the analyst who must solve the general PILP, this could be disastrous, given a limited computer budget. In the l.s.a., there is no way to monitor or control the amount of computer time spent on a particular (P_k). In fact, after an inordinate amount of computer time has been used, it is possible that only minimal progress may have been realized in the solution of each (P_k).

Furthermore, we point out that the primary advantages of the l.s.a. can be overcome by clever implementation of a serial approach. Two advantages of the l.s.a. are the small amount of bookkeeping calculations and core storage required. However, for more difficult PILPs, the bookkeeping time becomes insignificant, since the time required for solving the primary relaxation at each node generally is the lion's share of total computation time This was manifested in the facility location problem of Chapter VII, where over 90 percent of total computation time was devoted to the solution of transportation problems only. For more difficult problems, computer core limitations impose serious restraints on the serial approach. This is avoided in the l.s.a. However, by using an interactive serial approach (i.s.a.) as outlined in Chapter VII, node information is stored either on high speed disk or tape, and hence the core storage problem is alleviated. A third advantage of l.s.a. is the reoptimization capability from problem to problem in the PILP. That is, at a given node the optimal bounding problem relaxation solution for say, (CP_1), is used as a starting point for the solution of the relaxation for (CP_2). Oftentimes, this reoptimization technique can realize significant savings in computation time. This scheme may be used in the serial approach as well, by simply storing the optimal basis (when LP is used as the primary relaxation) for the relaxation of (CP_1) on disk or tape. Then, when the relaxation of (CP_2) is to be solved, the corresponding basis can be retrieved and used as the starting basis. Of course, this retrieval incurs a setup time, but, as pointed out earlier, setup time is insignificant for more difficult problems. Thus, the advantages of the l.s.a. can be neutralized by judicious use of an i.s.a.

We now analyze the i.s.a. and point out the more important factors in the implementation of such a method. By incorporating the human factor into the process, not only is closer control in monitoring the computation process possible, but greater flexibility in solution strategies can be realized. An i.s.a. allows the analyst to glean information from the solution process for (P_1) in order to solve (P_2) more efficiently. There are at least four sources of this information. They are:

1. An optimal solution for (P_1)

2. Root node penalties and a branch and bound tree for (P_1)

3. Success of branching criteria in the branch and bound tree

4. Relative effectiveness of various fathoming tests.

Each of these sources can be most helpful in planning a solution strategy for (P_2). In the following few paragraphs, we will analyze each of these sources.

Various feasibility recovery techniques have been presented for the special problem classes of chapters V, VI, and VII. For the general PILP, such techniques may or may not apply, depending on the type of parameterization. Therefore, to be assured of flexibility in the feasibility recovery technique, it may be advisable to have the analyst apply ad hoc techniques for obtaining a feasible solution to (P_2) from an optimal (or just feasible) solution to (P_1). Such techniques, of course, can be coupled with a computerized optimization scheme, for example, where all integer variables are fixed to a specific value.

The root node penalties and the branch and bound tree generated in solving (P_1) can be used as a guide in generating an initial separation for (P_2). First, consider the penalties generated at the root node for (P_1). Such penalties are generally calculated in order to choose branch variables. They may be LP based, or they may be a by-product of a Lagrangean relaxation. Lagrangean penalties can be quite a bit stronger and, at the same time, require fewer calculations [Geoffrion, 1974a]. By ranking the absolute values of these penalties in descending order, one can get some feeling for the "importance" of the individual integer variables. Such a ranking is valuable since in a branch and bound process it is preferable to branch on variables that have large penalties, early in the branching scheme. After (P_1) has been solved and the root node penalties have been ordered for (P_1), the root node penalties for (P_2) are calculated and ordered also. By comparing the orderings for (P_1) and (P_2), one can get some indication of whether the "important" variables in (P_1) are "important" in (P_2) also. If this is the case, it is reasonable to assume that the branch and bound tree for (P_1) will generate a "good" initial separation for (P_2). If, on the other hand, the orderings are quite different, the branch and bound tree for (P_1) may generate a relatively poor initial separation for (P_2). In this case, it may be better to either solve (P_2) with no initial separation or to use some separation based on the root node penalties for (P_2) with little or no input from the (P_1) solution process.

Considering the case where the root node penalty orderings are reasonably correlated, a number of possible choices exist for the generation of an initial separation. The simplest, of course, is to use the frontier of fathomed nodes from the branch and bound tree for (P_1). However, as we have seen in chapters V and VII such a separation may be rather large. A number of modifications can be made that reduce the size of the initial separation, while still retaining the power of the separation. First, variables that were pegged to a specific value can be "collapsed out" of the tree (compare Chapter VII, Section B). Second, branches that would not have been made if the optimal solution to (P_1) had been known at the beginning of the solution process can be eliminated. These additional branches can occur if the optimal solution value for (P_1) was not found until late in the branching process. In this case, the additional branches were made because the fathoming by value test used an inferior incumbent value. Selected branches may also be eliminated if the parent node of the branches had a relaxation value "close" to the optimal solution value for (P_1). Thus the node was "almost" fathomed, but since it just missed being fathomed, a branch had to be made. Since our general supposition in the serial method is that if a node is fathomed for (P_1), it will probably be fathomed for (P_2) as well, we see that it is reasonable to include nodes that are "almost" fathomed in our collection (and to eliminate the successor nodes of these nodes from the collection). Still another method for reducing the size of the initial separation is to use only those branches that are on the optimal path. That is, the path from the root node to the node where the optimal solution was found (see figure below).

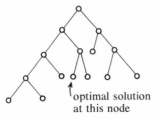

optimal solution
at this node

Branch and bound tree for (P_1)

Initial separation using
the optimal path only

For the case where the root node penalty orderings for (P_1) and (P_2) are not well correlated, another initial separation based on the (P_2) root node penalties can be generated. Specifically, we may commit a number of variables as branch variables using the (P_2) penalty ordering (see figure below).

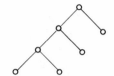

Initial separation for (P_2)
using root node penalty ordering

In this case, the left-hand branch constrains the variable to its "favorable" value as deduced from the penalties. The variable with the largest absolute penalty is committed first, the second largest second, and so on.

The relative effectiveness of branching criteria for (P_1) may be used in formulating the branching criteria for (P_2). A measure of the effectiveness of a branching criterion is the number of "mistakes" made in the branch and bound process. A *mistake* is defined to be investigation of the $x_i = 1$ branch first, say, when the $x_i = 0$ branch contains an optimal solution for (P_1). That is, the branching criterion indicates that the $x_i = 1$ branch is preferred over the $x_i = 0$ branch, when, in fact, the $x_i = 0$ branch contains an optimal solution. By keeping track of the mistakes in a branch and bound tree, one can identify a threshold penalty value below which the branching criterion may no longer be reliable. Furthermore, it may be possible to identify specific variables for which the branching criterion is unreliable. By categorizing variables according to their penalty indicators for (P_1) and (P_2) as well as according to the anticipated effect of the parameterization on each variable, one can predict the reliability of the (P_2) penalty indicator for each variable. The following display delineates the possible cases. In using this display, we assume that an optimal solution to (P_1) has been found, that the root node penalties have been calculated for (P_1) and (P_2), and that an intuitive analysis has been made concerning the effect of the parameterization on each variable. Further, we assume that all integer variables are 0–1.

Penalty indicator for (P_1)	Penalty indicator for (P_2)	Reliability of penalty indicator for (P_2)
Indicator agrees with optimal solution value in (P_1)	Indicator agrees with penalty indicator for (P_1)	a) Very reliable if parameterization tends to keep variable at same value b) Reliable if opposite of a

Indicator agrees with optimal solution value in (P_1)	Indicator disagrees with penalty indicator for (P_1)	c) Uncertain but tending to be reliable if parameterization tends to change value of variable d) Very uncertain if opposite of c
Indicator disagrees with optimal solution value in (P_1)	Indicator agrees with penalty indicator for (P_1)	e) Uncertain but tending to be reliable if parameterization tends to change value of variable f) Very uncertain if opposite of e
Indicator disagrees with optimal solution value in (P_1)	Indicator disagrees with penalty indicator for (P_1)	g) Uncertain but tending to be reliable if parameterization tends to keep variable at same value h) Very uncertain if opposite of g

By categorizing each variable by its reliability, one is able to modify the branching criterion in at least two ways. First, a threshold for variables can be established below which variables should be chosen for branching by some other criterion. Second, if a reliable variable and an uncertain variable are considered for becoming the next branch variable, the reliable variable should always be chosen. Thus, it is possible to modify branching criteria through the use of root node penalty indicators for (P_1) and (P_2), as well as by intuitive tendencies deduced from the type of parameterization.

The fourth source of information for making the (P_2) solution process more efficient is the effectiveness of various fathoming tests. For example, if some conditional logical tests were ineffective for (P_1), then one might consider the elimination of such tests for (P_2). Similarly, if calculations that attempt to tighten the primary relaxation generally fail for (P_1), they might be eliminated for (P_2).

In conclusion, the i.s.a. is a flexible approach that allows the analyst to modify solution techniques in such a way that the experience gleaned in solving (P_1) is used to maximum advantage in the solution procedure for (P_2).

B. Future Research

In this section, we give a brief outline of topics in the PILP area that are fertile for further research.

One such area is the study of the optimal solution as a function of a specific parameterization. We have alluded to this in chapters II and III. Hopefully, stronger and more useful results can be developed for other problem classes that take advantage of the knowledge of an optimal solution for one problem in order to solve a closely related problem. Such properties as continuity, monotonicity, and convexity of the optimal solution (possibly in certain components) and of the optimal solution value would seem to be worth seeking.

Still another area is the use of other methods such as cutting planes and group theory either alone or in conjunction with a branch and bound approach. We have outlined a rudimentary cutting-plane approach in Chapter I. It may be that such an approach would be quite effective. More promising, however, might be the incorporation of a cutting-plane capability in a branch and bound scheme.

Third, a parametric capability must be designed for use in commercial ILP codes. An interactive serial method as outlined in Section A of this chapter would seem to permit relatively simple and inexpensive implementation.

Fourth, it would seem likely that the solution method of Chapter V, Section D, can be extended to certain nonlinear parameterizations where the optimal objective value is a concave (for minimization) function of θ.

Fifth, the use of various data structures and list processing languages may be effective in coping with the data handling problem inherent in the PILP. Although list processing techniques were incorporated in the computer codes developed in this work, it is possible that improvements and efficiencies could be realized with the introduction of new programming techniques and languages.

Sixth, we note that techniques for parallelizing computations would be most effective in a PILP context. Such an approach would rely on the implementation of specific computer data structures. Furthermore, the advent of fourth generation computers such as ILLIAC IV that perform computations in parallel should also have a telling impact on PILP algorithms.

In conclusion, we surmise that parametric methods in integer programming will assume greater importance as ILP solution methods improve. Just as in LP, where parametric analysis has become an expected and useful part of most solution studies, we expect the same to occur in PILP. We acknowledge that our study of PILP is just a beginning. The true test of a method lies in its use by those who can benefit by its availability.

Bibliography

Akinc, U., and Khumawala, B. M. "An Efficient Branch and Bound Algorithm for the Capacitated Warehouse Location Problem." *Management Science* 23:6 (February 1977): 585–94.

Benichou, M.; Gauthier, J. M.; Girodet, P.; Hentgès, B.; Ribière, G.; and Vincent, O. "Experiments in Mixed Integer Linear Programming." *Mathematical Programming* 1:1 (1971): 76–94.

Bowman, V. J. "The Structure of Integer Programs under the Hermitian Normal Form." *Operations Research* 22:5 (September–October 1974): 1067–80.

———. "Sensitivity Analysis in Linear Integer Programming." *AIIE Technical Papers* (1972).

DeAngelo, H. C., private communication (March 1974).

Dembo, R. S., private communication (May 1974).

Efroymson, M. A., and Ray, T. L. "A Branch-and-Bound Algorithm for Plant Location." *Operations Research* 14:3 (May–June 1966): 361–68.

Ellwein, L. B. "Fixed Charge Location-Allocation Problems with Capacity and Configuration Constraints." Technical Report No. 70-2, Department of Industrial Engineering, Stanford University, California, August 1970.

El-Shafei, A. N., and Haley, K. B. "Facilities Location: Some Foundations, Methods of Solution, Applications, and Computational Experience." Operational Research Report No. 91, North Carolina State University at Raleigh, May 1974.

Garfinkel, R. S., and Nemhauser, G. L. "The Set Partitioning Problem: Set Covering with Equality Constraints." *Operations Research* 17:5 (September–October 1969): 848–56.

———. *Integer Programming.* New York: John Wiley and Sons, Inc., 1972.

Geoffrion, A. M. "Lagrangean Relaxation for Integer Programming." *Mathematical Programming Study 2: Approaches to Integer Programming*, 1974a, pp. 82–114.

———. "Distribution Systems Configuration Planning: A Strategy for Managerial Decisions Through Computer-Based Analysis." Western Management Science Institute Working Paper No. 219, University of California, Los Angeles, September 1974b.

———. "A Guide to Computer Assisted Methods for Distribution Systems Planning." *Sloan Management Review* (1975).

Geoffrion, A. M., and Graves, G. W. "Multicommodity Distribution System Design by Benders Decomposition." *Management Science* 20:5 (January 1974c): 822–44.

Geoffrion, A. M., and Marsten, R. E. "Integer Programming Algorithms: A Framework and State-of-the-Art Survey." *Management Science* 18:9 (May 1972): 465–91.

Geoffrion, A. M., and Nauss, R. M. "Parametric and Postoptimality Analysis in Integer Linear Programming." *Management Science* 23:5 (January 1977): 453–66.

Greenberg, H., and Hegerich, R. "A Branch Search Algorithm for the Knapsack Problem." *Management Science* 16:5 (January 1970): 327–32.

Hammer, P. L., and Nguyen, S. "A Partial Order in the Solution Space of Bivalent Programs." Presented at the 41st Meeting of ORSA, New Orleans, Louisiana, April 1972.

Horowitz, E., and Sahni, S. "Computing Partitions with Applications to the Knapsack Problem." *Journal of the Association for Computing Machinery* 21:2 (April 1974): 277–92.

Klein, D., and Holm, S. "Integer Programming Postoptimal Analysis with Cutting Planes." Working Paper, Institute of History and Social Science, Odense University, Odense, Denmark, April 1977.

Korsh, J. F., and Ingaragiola, G. P. "Reduction Algorithm for Zero-One Single Knapsack Problems." *Management Science* 20:4 (December 1973): 460–63.

Kuehn, A. A., and Hamburger, M. J. "A Heuristic Program for Locating Warehouses." *Management Science* 9:9 (July 1963): 643–66.

Manne, A. S., ed. *Investments for Capacity Expansion: Size, Location, and Time-Phasing*, p. 201. Cambridge, Mass.: The M.I.T. Press, 1967.

Marsten, R. E., *An Implicit Enumeration Algorithm for the Set Partitioning Problem with Side Constraints*. Ph.D. diss., University of California, Los Angeles, October 1971.

Marsten, R. E., and Morin, T. L. "Parametric Integer Programming: The Right Hand Side Case." Working Paper, Sloan School of Management. Cambridge, Mass.: The M.I.T. Press, July 1975.

Martin, G. T. "An Accelerated Euclidean Algorithm for Integer Linear Programming," in R. L. Graves and P. Wolfe, eds., *Recent Advances in Mathematical Programming*. New York: McGraw-Hill, 1963.

Noltemeier, H. "Sensitivitalsanalyse bei disketen linearen Optimierungsproblemen," in M. Beckmann and H. P. Kunzi, eds., *Lecture Notes in Operations Research and Mathematical Systems*, No. 30. Springer-Verlag, New York, 1970.

Piper, C. J., and Zoltners, A. A. "Implicit Enumeration Based Algorithms for Postoptimizing Zero-One Programs." Management Sciences Research Report No. 313, Graduate School of Industrial Administration, Carnegie-Mellon University, March 1973.

———. "Some Easy Postoptimality Analysis for Zero-One Programming." *Management Science* 22:7 (March 1976): 759–65.

Radke, M. A. "Sensitivity Analysis in Discrete Optimization." Working Paper No. 240, Western Management Science Institute, University of California, Los Angeles, September 1975.

Roodman, G. M. "Postoptimality Analysis in Zero-One Programming by Implicit Enumeration." *Naval Research Logistics Quarterly* 19 (1972): 435–47.

———. "Postoptimality Analysis in Integer Programming by Implicit Enumeration: The Mixed Integer Case." *Naval Research Logistics Quarterly* 21:4 (1974): 595–607.

Ross, G. T., and Soland, R. M. "A Branch and Bound Algorithm for the Generalized Assignment Problem." *Mathematical Programming* 8:1 (February 1975): 91–103.

Sà, G. "Branch-and-Bound and Approximate Solutions to the Capacitated Plant-Location Problem." *Operations Research* 17:6 (November–December 1969): 1005–16.

Wagner, H. M. *Principles of Operations Research*, Englewood Cliffs, N. J.: Prentice-Hall, Inc., 1969.

Index